CONTENTS

ILLUSTRATIONS

TABLES

GLOSSARY

Aerostat	A reusable, inflated, lighter-than-air platform in this context
APOD	Aerial Port of Debarkation
ASL	An altitude Above Sea Level
Balloon	A disposable, inflated, lighter-than-air bag in this context
CBI	China-Burma-India Theater of WWII combat operations
CCAAF	Clinton County Army Air Field in Wilmington Ohio
CG	Cargo Glider is a United States Army Air Force designation.
CSS	Combat Service Support is one of three USMC combat elements.
GVW	Gross Vehicle Weight, a vehicle's design maximum in this context
HA/DR	Humanitarian Assistance and Disaster Relief
HMMWV	High Mobility Multi-purpose Wheeled Vehicle
iHL	Interim Heavy airlift system
ISO	International Organization for Standardization
JATO	Jet-Assisted Takeoff
JMIC	Joint Modular Intermodal Container; a 52x44x42-inch packaging standard
JOA	Joint Operating Area
LAV	Light Armored Vehicle
LCAC	Landing Craft Air Cushion
LW155	M777A1 Lightweight 155mm howitzer
LZ	Landing Zone; typically unimproved, soft, or uneven land used by air vehicles
MEB	Marine Expeditionary Brigade
MTVR	Medium Tactical Vehicle Replacement
MPF (F)	Maritime Prepositioning Force (Future); a squadron of 14 ships
NSWCDD	Naval Surface Warfare Center Dahlgren Division
ONR	Office of Naval Research
PHST	Packaging, Handling, Shipping, and Transport
Phugoid	Aircraft motion of pitching up and then back
Rotation	Lifting flight upon accelerating to a velocity greater than stall speed
SBME	Sea Base Maneuver Element
Sea base	Ships projecting and sustaining warfighters ashore
STARS	Surface-To-Air-Recovery System
STOL	Short Takeoff and Landing
STOM	Ship-to-Objective Maneuver
TEU	Twenty-foot Equivalent Unit; an 8x8.5x20-ft shipping container capacity
TTP	Tactics, Techniques, and Procedures
USMC	United States Marine Corps
USN	United States Navy
VERTREP	Vertical Replenishment; rotorcraft-based, at-sea transit
V-J Day	Victory over Japan
Volplane	To glide to the earth; in this context after a standoff release
WWII	World War II
XG	Experimental logistics glider designation

EXECUTIVE SUMMARY

This technical report proposes an interim heavy airlift (iHL) concept to connect the littoral Sea base to warfighters ashore. Ships capable of selective offload use the iHL system to provide materiel over the horizon by distributed air means to Ship-to-Objective Maneuver (STOM) forces moving rapidly to operational objectives without stopping to seize, defend, and build up beachheads or landing zones. iHL purposely does not alter planned Navy Sea base constructions. iHL scales to sustain any force ashore from small units and Distributed Operations up to Marine Expeditionary Brigade (MEB) operations. iHL can reduce the number of rotorcraft or tilt-rotorcraft it would otherwise take to daily resupply by air, and uses ship cargo space for storage.

iHL rediscovers proven US military air transport concepts in a novel combination during its key performance stage upon the Sea base—helipad snatch pickup. Existing aircraft intercept a balloon in a snatch pickup of a logistics glider for multiple-towed transport in austere delivery to tactical maneuvering units. A fixed-wing logistics glider is proposed to consume exact multiples of International Organization for Standardization (ISO) standard volumes aboard the Sea base for storage efficiency.

This report explores a range of notional vehicles within representative models of the proposed system for a design trade space. It is shown that iHL is a viable concept worth pursuing. Verification of this modeling is advised, first by software simulation and then by advanced concept demonstration.

The historical challenges associated with these US military air transport concepts have long since been overcome and operationally proven. While its aviation aspects might be near an atypical flight envelope to the modern Navy, iHL is dominated by standardized packaging, handling, shipping, and transport (PHST) of both the cargo payload and the new air vehicle. Logistics gliders require both a capable supply chain and an effectively trained end user. The air community – including joint Navy, Air Force, Army, Coast Guard and Coalition forces – must be involved in the aviation aspects of iHL, but the expeditionary surface and ground community must make the difficult tradeoff decisions on how the expeditionary logistician will operate iHL.

1 INTRODUCTION

This technical report proposes an interim heavy airlift (iHL) concept to connect the littoral Sea base to warfighters ashore. Ships capable of selective offload use the iHL system to provide materiel over the horizon by distributed air means to Ship-to-Objective Maneuver (STOM) forces moving rapidly to operational objectives without stopping to seize, defend, and build up beachheads or landing zones.[1] iHL purposely does not alter planned Navy Sea base constructions. iHL scales to sustain any force ashore from small units and Distributed Operations up to Marine Expeditionary Brigade (MEB) operations. iHL can reduce the number of rotorcraft or tilt-rotorcraft it would otherwise take to daily resupply by air, and uses ship cargo space for storage.

1.1 Sea Base Airlift

NSWCDD analysis shows different models dominate performance at each link in the expeditionary resupply chain.[2] Timely and precise delivery of materiel is the key. For ocean distances, the measure of success is a significant volume of materiel maintaining a medium speed of advance in the higher sea states. For the shorter littoral distances, speed of advance can be low, but not zero, and is dependent upon sea state, hostile action, and success at each supply chain link. Timeliness in delivery is sensitive to the number and availability of littoral and ground handling links, and to low ground speed over medium distances. Shore depots provide a buffer against threats but are imprecise for delivery. Precision aerial resupply from the Sea base bypasses these littoral and ashore links at a high speed of advance, making the concept effective and attractive. The issue then becomes defining economical means to create this capability.

The unavailability of forward bases, ports, and airports increases the value of a Sea base. However, Sea base deck space and the lift to dedicate to all-aerial resupply are currently unaffordable while lift vehicles have limited range. Self-launch of significant-capacity lift is not yet viable, given the limits of the planned Sea base constructions. However, assisted launch in quantity can be achieved with the proposed reusable technique for self-lifting flight using many, smaller capacity air cargo vehicles.

iHL rediscovers proven US military air transport concepts in a novel combination during its key performance stage upon the Sea base—helipad snatch pickup. Existing aircraft intercept a balloon in a snatch pickup of a logistics glider for multiple-towed transport in austere delivery to tactical maneuvering units. The fixed-wing logistics glider is proposed to consume exact multiples of ISO standard volumes aboard the Sea base as cargo itself for storage and handling efficiency. The fixed wing structure is low maintenance yet supports heavy loads; and a cargo glider's high useable payload percentage of gross vehicle weight (GVW) requires less Sea base-supplied resources or a smaller "footprint" in launching each ton of materiel.

The WWII challenges in aviation technology, landing zone (LZ) surveys, and operational logistics have long since been overcome and operationally proven. While its aviation design incorporates modern technologies, iHL is a novel link between expeditionary supply chains. Performance via packaging, handling, shipping, and transport (PHST) dominate system effectiveness. The air community – including joint Navy, Air Force, Army, Coast Guard, and Coalition forces – must be involved in the aviation aspects of iHL, but the expeditionary surface and ground community must make the difficult tradeoff decisions on how the expeditionary logistician will operate iHL

This technical report presents and models the iHL concept by the systems engineering of expeditionary logistics. Its potential trade space is explored, and a short developmental path to concept demonstration is prescribed. Chapter 2 starts with a comprehensive logistics systems engineering perspective for the historical pedigree of iHL. This establishes a baseline context of experiences and characteristics. Next the iHL concept is introduced with key decisions to further explore, such as reuse, occupied flight, and autonomous control. Several notional logistics glider models frame the applicable trade space. Then performance modeling of iHL sustainment of the Sea Base Maneuver Element (SBME) MEB ashore indicates iHL is realistically viable. Chapter 3 mitigates risk by first modeling the physical forces from WWII measurements as a baseline. Helipad snatch pickup is found to be within previously demonstrated capability and can extend significantly beyond it. Chapter 4 proposes a development plan to achieve an iHL demonstration.

The concluding chapter recommends the modeling and simulation of helipad snatch initially to verify what was shown here to be viable. This has significant influence upon such issues as useable vehicle weight for the design tradeoff decisions. This then initiates the presentation of iHL to its diverse developmental and operational community.

1.2 Supply and Standardization

Cargo gliders faded in the post-WWII production slump. Helicopters combined with Cold War basing infrastructure to effectively overcome their individual shortcomings. Consequently, there was little need for the longer supply chains and flexibility of a Sea base to resupply expeditionary forces.

The Navy last standardized logistics processing with the pallet in 1958. The Marine Corps standardized on the Twenty-foot Equivalent Unit (TEU) ISO packaging standard in 1974. The commercial shipping and ground transportation industry has embraced TEU standards for efficiency in handling and packaging in shipping, storage, and transport. The new Joint Modular Intermodal Container (JMIC) specification holds the promise of dense containment within the TEU and efficiency during the repackaging links of the Sea base supply chain. There is no apparent standardization on naval expeditionary repair and consumables such as tires.

The performance demands placed upon the Sea base require the most effective use of cargo space and processing resources. With standardized packaging, it is proposed that the iHL footprint upon the Sea base be multiples of the TEU volume and stored as selective offload cargo. With the delivery vehicle processed as cargo, the constrained Sea base gets ideal performance from the iHL delivery concept.

1.3 Non-Military Applications

The number, frequency, and scale of humanitarian assistance and disaster relief (HA/DR) operations can be expected to increase in the future. Their economic preparation and effective delivery will benefit from a standardized container infrastructure. Since their destinations are not known before disaster strikes, assistance packages should be ready in ports for surge operation, or ideally, already aboard prepositioned supply ships.

Since a high percentage of the human population lives near littoral waters, it is expected that a disproportionately high percentage of HA/DR operations will be within reach of the Sea base. For in-theater arrival and distribution, utilizing the cargo delivery vehicle out of the cargo is the densest packing for economic storage and transport and is optimal for effective handling and delivery. The air transport of standardized containers is ideal for timely delivery to regions that are remote, austere, or unexpectedly limited in access. iHL proposes a marriage of these two ideals.

To name a few standardized container infrastructure examples, this may include the following facilities:
- Housing
- Storage
- Power
- Water treatment
- Sanitation
- Medical
- Command and control
- Communications
- Incarceration

Beyond their standardized nature, these non-military applications are not explored further in this report.

2 CONCEPT

US WWII cargo gliders are best remembered for their legendary role in invasion delivery—and those aboard have earned respect in military history. Opinions on the overall effectiveness of the cargo glider system cover a range of emotions, however, and in contrast to the invasion function, the system's performance in austere transport was unrecognized until now. Cargo gliders were a multiplier to air cargo transport, but also had a relevant austere logistical transport capability with its recovery technique. This performance record is examined using modern systems engineering techniques, and new insights are gained from the perspective of glider recovery and its influence upon large cargo gliders.

These salient historical operational experiences, performance characteristics, and systems interactions are selected as the basis from which iHL expands. Then this chapter introduces the logistics glider flight profile, and key design tradeoffs for implementation decisions are discussed. Next, notional logistics glider models are sufficiently detailed for the ensuing analysis of daily resupply. The chapter concludes with all possible iHL tactical actions and expands into stages of the iHL tactical lifecycle, including options for advancing technology or speculative capabilities inserted into the iHL system.

2.1 Proven US Military Concepts

This section describes the baseline model under consideration. Each stage of tactical iHL operation is modeled upon relevant aspects of previous US military air transport systems; however, those systems have not been used recently and not ever together. Their rediscovery and reapplication is combined with modern sciences and new technologies for a viable system.

2.1.1 Snatch Pickup

The history of cargo pickup-on-the-move dates to the early railroad days, when trains moving through remote towns snatched hanging mailbags. Originally this may have been because the engines did not have enough torque to start up again on non-level rails, since not every town had a level location for a station. Later, it was a time-saver for freight trains not to stop and start up again at each mail spot along their routes.

The Marines first demonstrated aerial snatch pickup with leather dispatch bags in 1927 using a surplus World War I DH-4 biplane.[3] This technique was applied to rural airmail pickup and delivery in the 1930's by the All American Aviation Company thanks to partnering by Richard Chichester duPont. In 1941, this team developed glider snatch using DuPont Corporation's new undrawn nylon towline[4] (uninylon[5]). Escalating through heavier sailplanes, this technique transitioned in 1942 at two secret test and experimentation facilities near Dayton, Ohio, for Army Air Corps post-invasion cargo glider recovery. Two former All American Aviation civilians and a then-Army Air Corps

Captain Lee Jett were the snatch pickup test pilots of a great team who refined the technique by experimentation.

Figure 1 is an in-theater photograph of glider snatch at the moment of intercept.[6] Training film TF-1-3399 explains in detail glider snatch procedures involving the glider, towline, ground intercept station, and tow plane.[7] The tow craft is also called a "tug."

Photo Removed Due to Copyright Restrictions

Figure 1. First Normandy Pickup (Photo by Yves Tariel of Paris France)

The reusable towline is the key to the entire system and its parts count is surprisingly high. The details within Figure 2 show a sophisticated combination of metal and nylon swivels, weak links, and lines of varying diameters.[3]

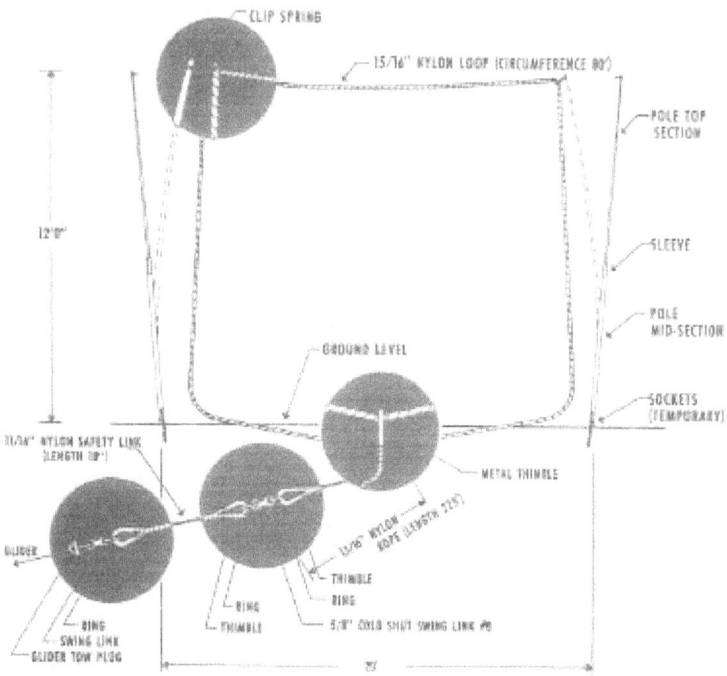

Figure 2. Glider Pickup Ground Station Unit

These diagrams describe the snatch pickup method. The fully assembled ground intercept station of Figure 2 has two station poles safely offset from the glider's launch path, standing with the glider's towline looped tightly between them.[8] On an unimproved field, road, or runway, the glider per Figure 3 is at an angle to the airplane's approach track.[9] The tug has a pickup arm to direct the intercept of the towline into a hook at the end of a steel cable. In Figure 4, the tug swoops low and intercepts the ground station.[8] A winch onboard the tug pays out the steel cable for several hundred feet, gradually engaging a pre-set clutch to increase towline tension. The nylon towline elongates under the load to absorb the inertial differences. The tug path follows an arc and it pulls up. As the winch locks, the glider accelerates from 0 to tow speeds over 90 mph in 6 or 7 seconds.

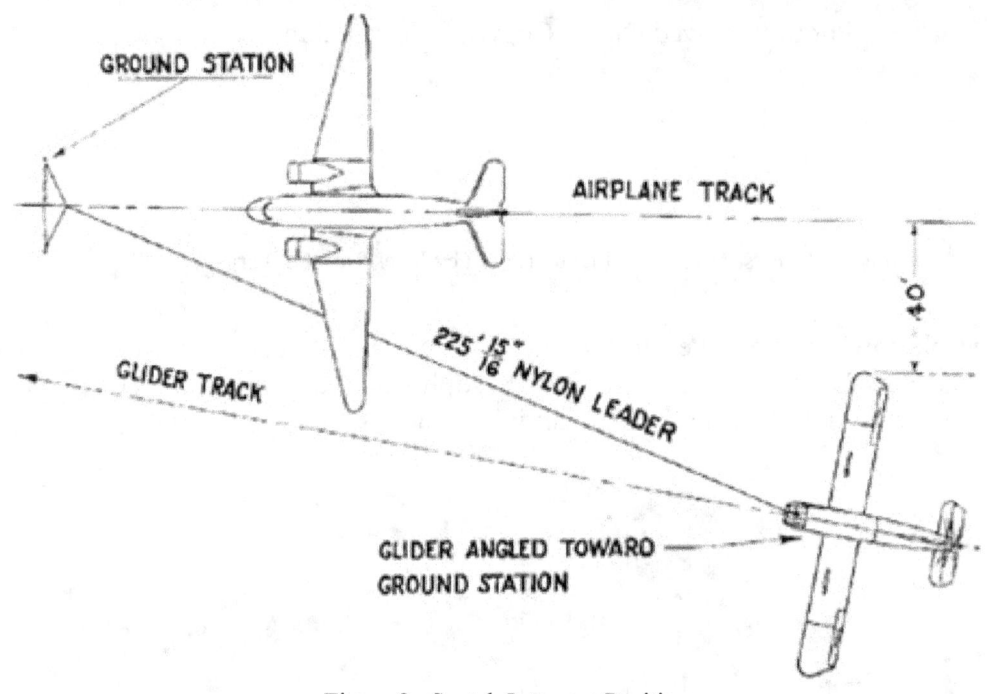

Figure 3. Snatch Intercept Positions

The airplane track is into the wind. The glider is offset to the airplane track to avoid contact with the low flying tug and its boom, and to avoid running over the ground station during takeoff.

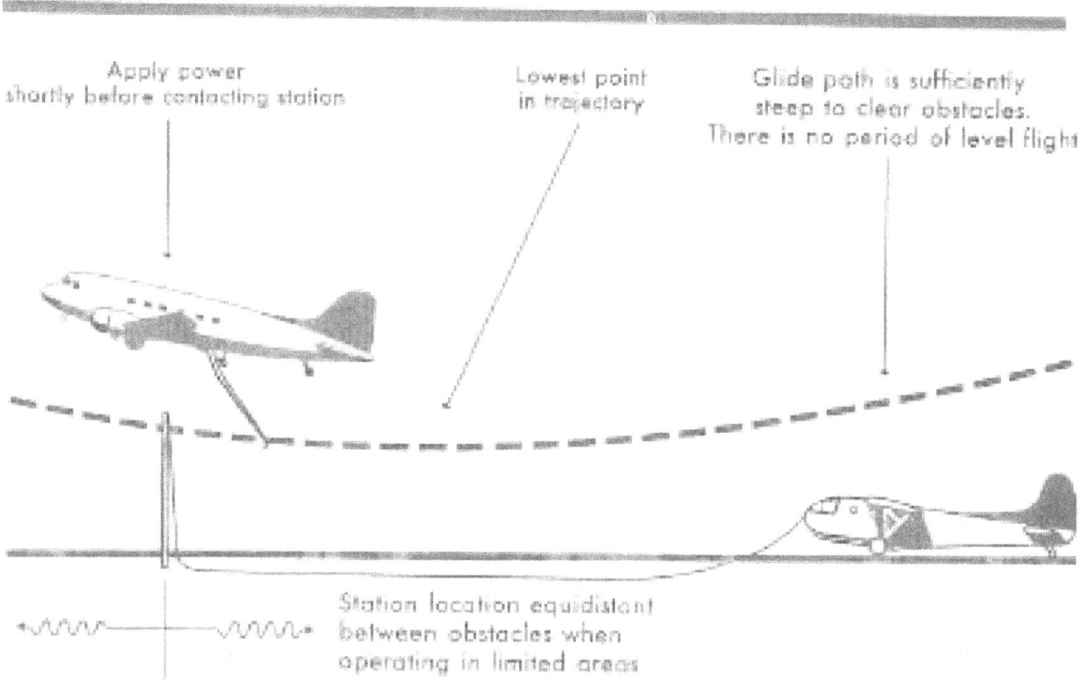

Figure 4. Tug Intercept Trajectory

The tug's climb upwards gets the glider airborne quickly so they clear ground-based obstacles. Clearing 50-foot obstacles was their primary motivation rather than distance spent on the ground.

The glider can climb faster than the tug. Lee Jett describes one training incident at Clinton County Army Air Field (CCAAF)[10] in which an inexperienced glider pilot nosed his glider too high during snatch climb out. The steel cable contacted and momentarily raised the tug's elevator. The elevator fabric was damaged and had to be replaced. A pushbutton-activated pyrotechnic was then devised for emergency towline separation.[11] This feature transitioned into the production system.[7]

Late in 1942, contracts were let for pickup equipment for heavier glider snatches in the 8,000- to 16,000-lb range.[10] Starting with the Model 80 unit, these contracts produced a series of winches that raised cargo gliders' pickup capacities to 25,000 lbs at CCAAF.[10] Table 1 lists the tug winch pickup unit product line as of 1947.[12] The largest unit developed is understood to be the Model 160; the Model 200 was only proposed.[13]

Table 1. All American Aviation Heavy Duty Pickup Units

UNIT	MAX. GLIDER WEIGHT	SPEED	WT. OF UNIT	CABLE SIZE	CABLE LENGTH	CABLE BREAKING STRENGTH	DIMENSIONS			K.E. CAPACITY
							WIDTH	HEIGHT	FORE & AFT	
Model 4	550 lbs. 250 kg	70 m.p.h 110 km/h	95 lbs. 43 kg	1/8 in. 3.18 mm	700 ft. 213 m	2000 lbs. 910 kg	15 in.* 394 mm*	22½ in.* 570 mm*	19½ in.* 495 mm*	100,000 ft-lb. 13,800 m-kg
Model 15	1500 lbs. 680 kg	120 m.p.h. 195 km/h	144 lbs. 65 kg	1/8 in. 3.18 mm	1000 ft. 305 m	2000 lbs. 910 kg	23 in. 585 mm	24 in. 610 mm	24 in. 610 mm	720,000 ft-lb. 100,000 m-kg
Model 40	4000 lbs. 1800 kg	120 m.p.h. 195 km/h	275 lbs. 125 kg	1/4 in. 6.35 mm	1000 ft. 305 m	7000 lbs. 3200 kg	27½ in. 700 mm	26 in. 660 mm	26 in. 660 mm	1,930,000 ft-lb. 265,000 m-kg
Model 80	8000 lbs. 3600 kg	120 m.p.h. 195 km/h	620 lbs. 280 kg	3/8 in. 9.5 mm	1080 ft. 330 m	14,400 lbs. 6500 kg	40 in. 1020 mm	31½ in. 800 mm	30 in. 760 mm	3,610,000 ft-lb. 500,000 m-kg
Model 81	8000 lbs. 3600 kg	135 m.p.h. 215 km/h	778 lbs. 355 kg	3/8 in. 9.5 mm	1080 ft. 330 m	14,400 lbs. 6500 kg	33 in. 840 mm	29½ in. 750 mm	30 in. 760 mm	5,000,000 ft-lb. 690,000 m-kg
Model 120	12,000 lbs. 5450 kg	140 m.p.h. 225 km/h	1250 lbs. 565 kg	1/2 in. 12.7 mm	1175 ft. 360 m	22,800 lbs. 10,300 kg	44 in. 1120 mm	40 in. 1020 mm	39 in. 990 mm	7,710,000 ft-lb. 1,065,000 m-kg
Model 160	16,000 lbs. 7250 kg	140 m.p.h. 225 km/h	1900 lbs. 860 kg	9/16 in. 14.3 mm	1350 ft. 410 m	35,000 lbs. 16,000 kg	45 in. 1140 mm	36 in. 915 mm	52 in. 1320 mm	10,500,000 ft-lb. 1,450,000 m-kg
Model 200	20,000 lbs. 9000 kg	150 m.p.h. 240 km/h	3100 lbs. 1400 kg	9/16 in. 14.3 mm	1350 ft. 410 m	35,000 lbs. 16,000 kg	49 in. 1245 mm	62½ in. 1590 mm	74½ in. 1890 mm	15,000,000 ft-lb. 2,070,000 m-kg

*Excluding Electric Motor.

Demonstration of runway takeoff with dual towed cargo gliders occurred first at Wright Field[11,14] and then at CCAAF.[15] This was rehearsed before Operation Neptune[16] and implemented in Operation Varsity.[15] It is still occasionally performed with modern recreational sailplanes.

Two gliders being sequentially snatched by a tug was first demonstrated in July 1942 at CCAAF.[4,10] The transfer of the towline off the winch between double snatch pickups was documented[3] and photographed.[17] It was performed twice by Lee Jett's expert crew between 1943 and 1946.[11] During double snatches, it was important for the glider pilot to maintain separation from the other glider while preventing the towline from interfering with the tug's elevator.[10]

At least three wartime factories used snatch pickup to deliver cargo gliders[10] for fastest receipt to government facilities and also not to involve wartime surface transport infrastructure. It was routine to snatch gliders from fields after towline breaks, typically during cross country transfers.[11] Stateside, Lee Jett alone performed approximately 2,500 cargo glider and non-glider snatch pickups[10] from 1942-1946.[11]

In the field, 474 cargo glider snatch pickups were documented across four theaters, and in 19 months followed half of the eight major combat glider missions. Table 2 is the first comprehensive snatch pickup list. The "Gliders" column lists the mission's effective glider count.

Table 2. Operational Snatch Pickups

Theater and Mission	Date	Gliders	Pickups
China-Burma-India[18]			
Exercise recoveries	9 Jan 1944	16	16
First snatch pickups behind enemy lines	29 Feb 1944	2	2
Operation Thursday, Burma	Mar 1944	68 [19]	0
Operation Capital medical evacuations	Oct 1944	25	25
Radar shipment	12 Feb 1945	5	5

Table 2. Operational Snatch Pickups (Continued)

Europe[19]			
Operation Husky, Sicily[16]	9 Jul 1943	136	0
Operation Neptune recoveries, Normandy France (possibly 15 – 40)	23-25 Jun 1944	517	13
Operation Dragoon, Southern France[6]	15 Aug 1944	407	*unknown*
Operation Market Garden recoveries, Holland (possibly 281)	Oct - Dec 1944, Feb 1945 [24]	1,900	256
Operation Repulse, Bastogne Belgium[6]	26-27 Dec 1944	61	0
Operation Varsity medical evacuations	22 Mar 1945	2	2
Operation Varsity recoveries, Germany	Apr- 1945	906	148
Pacific			
Operation Gypsy Task Force, Philippines[6]	23 Jun 1945	7	*unknown*
"Shangri La" valley rescue, New Guinea[10,20]	2 Jul 1945	3 [19]	3
Arctic			
Alaska rescue[21]	14 Dec 1948	1	1
Greenland icecap rescue attempts[22]	17,25 Dec 1948	2	3

Top crew and equipment were dispatched from Wright Field to North Africa for Sicily recoveries, but those gliders had since deteriorated.[15] Pickup quantities for Operations Dragoon[24] and Gypsy Task Force are unknown. It is unlikely there were recoveries in Operation Repulse.[19,24] That wholly successful combat resupply to Bastogne absorbed ground fire and, with the weather conditions, made the reuse of such a small force unlikely. Sources differ on the pickup tallies from Operations Neptune and Market Garden possibly due to bookkeeping errors, differentiation between US and Allied inventory, or used runways for towed recovery rather than snatch pickup.

The averaged snatch pickup rate from the table is nearly 12%, so roughly 1 out of 8 in-theater gliders was snatch recovered. The number and variety of snatch pickups surprises historians because it was discouraged on any significant scale for the European Theater[23] with winches removed upon arrival in England.[15] At least initially, dissatisfied with the returns after Operations Neptune and Dragoon, this decision was reversed for post-Operation Market Garden in the initial attempt at large-scale glider recovery. Unfortunately an October storm wrecked an additional 115 gliders earmarked for recovery.[3] Otherwise the overall snatch pickup total would have been at least 24% higher. The Battle of the Bulge suspended this effort for about two months,[24] and the salvage effort was completed over five months after the start of Operation Market Garden.

Except for two successful medical evacuations, all European Theater snatch pickups were post-invasion salvage. The famous "Shangri La" rescue used snatch pickup to extract crash survivors in the far inland jungle at 5,000 ft elevation near hostile territory. The remaining theaters of operation are summarized next.

2.1.1.1 China-Burma-India Operations.

The China-Burma-India (CBI) Theater demonstrated novel cargo glider applications in successful invasion, transport, and rescue operations.

The Army Air Corps had several special warfare groups before the official formation of the Air Force Special Operations Wings. Lee Jett helped train codename Project 9 tow pilots in glider snatch prior to their departure to CBI.[11] They became the 1st Air Commando Group, which included 150 cargo gliders[18] performing a series of successful disruptive actions starting with Operation Thursday on March 5, 1944.[25] This air unit transported and supplied the British coalition Chindit army in preventing the Japanese invasion of India by establishing a series of forward operating bases hundreds of miles behind enemy lines.

Training experimentation established a straight-in final approach from 200 yards out rather than the traditional four-leg landing pattern. In a preparatory exercise in January 1944, 16 gliders landed in an unexpectedly muddy LZ and were snatched out the following morning. Two gliders were recovered the next month as part of a successful covert insertion behind enemy lines. Although no snatch pickups were documented during the following six months of campaigning instigated by Operation Thursday, their glider section compiled impressive statistics moving brigades, battalions, and supplies in combat.[18]

It was common during conventional transport operations everywhere for the CG-4A model to gross around 9,000 lbs, or 38% beyond rated payload capacity. CBI towlines often failed when their dual-towed, significantly overloaded gliders surged simultaneously during descent over mountains. Nor did glider designers envision an unusual payload with airlifted armies: Thousands of pack animals were transported, including horses, mules, and bullocks. CBI casualties were typically evacuated by C-47's or light planes and even once by an R-4 helicopter. But in Operation Capital, two tugs towed four gliders to deliver 31,000 lbs of materiel, and in 25 snatches evacuated 123 casualties.[18]

The final documented CBI snatch pickups were at a shipping-receiving location. It was easiest to bring gliders to the cargo and then snatch them for delivery to a remote location for radar installation.[18]

2.1.1.2 Arctic Rescue Operations.

Postwar arctic rescue operations used cargo gliders and snatch pickup. CG-15A models had winterized conversions.[15] Snatch pickup was demonstrated on the (presumed frozen) Arctic Ocean[21] likely as part of a training exercise. There were two separate arctic rescue operations in December 1948. The Alaska pickup of the six-man crew of a downed transport was successful, while the Greenland ice cap pickups were not.

On Dec. 7, 1948, an Air Force C-47 crash-landed in the Greenland interior at 8,000 ft elevation without injury to the crew. A rescue B-17 crashed while landing. Next a C-54 delivered an arctic winterized glider. In 30 minutes its crew set up the station poles and towline for a snatch pickup. But the towline snapped just as the glider became airborne. A second snatch repeated the problem. High winds overnight destroyed the glider.[22]

On Christmas Day this failed again. Only this time the nose of the second glider was destroyed by towline whip back. The still-uninjured survivors and would-be rescuers were finally rescued after a total of three weeks by a ski-equipped C-47 with Jet Assisted Takeoff (JATO) rockets.[22]

Unlike the powered aircraft, there is no official accident report for the two gliders, so section 3.1.2.1, "Greenland Rescue Analysis," speculates on the cause of the towline failures.

2.1.2 <u>Glider Evolution to Snatch Pickup</u>

While hardly a motivation or even well understood during this era, systems engineering toward snatch recovery nevertheless had a significant influence on the evolution of the invasion glider into an austere transport system.

The CG-4A was the renowned WWII invasion glider. It was built by 16 different prime contractors[10] across the US. This model was intentionally low technology for non-aviation manufacturing industries[26] to convert to production on a large scale. Many saw the CG-4A as one-way delivery of Army infantry to unimproved landing zones, to be abandoned where they landed. While it was a low technology assembly with budget-conscious materials, the reality was much different than expected.

- The CG-4A had 70,000 parts.[6]
- Subcontracting for those parts proved problematic.[10]
- Many of the converted production industries failed to deliver useful quantities.[10]
- Targeted production cost never reached expectation[10] for disposable delivery.
- Assault operations were unexpectedly dangerous[16] for still maturing insertion tactics.
- The mounting for the towline was off-axis, inefficient in snatch pickup and hence any reuse.

A significant majority of cargo gliders did deliver successfully to unimproved LZs. However, there was not much of military significance recovered post-invasion given the successes of the various threat environments. For many reasons the high-volume European Theater failed in large-scale recovery.

- The invasion mindset did not contemplate reuse for the next major assault. Each was the last big one.
- The enemy had invented glider invasion and did employ effective countermeasures to the Allied secret weapon.
- There was a dearth of snatch training and equipments for air and ground crews.[11]
- Gliders were treated akin to trailers; they were not assigned call numbers and were referenced by model number.[21]
- There was a robust supply of fresh inventory.
- Gliders landed intact within tree-lined fields that prevented the snatch maneuver.[3]

11

- Components did not survive prolonged or harsh exposure to the elements and were scavenged by local residents.[10]
- There was little interest in recovery after Operation Varsity entered Germany.

Nonetheless glider snatch pickup did occur far more often and in more ways than expected. Follow-on large glider development emphasized survivability and capacity. Assets that survive get reused. Consequently greater pickup capacity developed in the last generation of cargo glider models as designers looked beyond the European Theater.

The Pacific Theater had less reliable supply lines by its topology. When compared to the Atlantic and overland supply lines of Europe, the Pacific island-hopping depots had transfer complexity, and hostile and sea threats. The Pacific campaign instituted the modern concept of naval logistics.[27] An end link to that supply chain was larger capacity gliders to replace or supplement the ubiquitous CG-4A. Table 3 lists, by increasing weight, the performance ranges for production cargo gliders—including the pair of XCG-10 prototypes. The speeds in parentheses are at full load without using flaps if available.

Table 3. Production Cargo Gliders by Weight

Model	Weight (lbs)	Wingspan (ft)	Wing area (sq ft)	Speed (mph)
CG-4A	3,500 - 7,500	83' 8"	852	41 (55) - 150
CG-15A	4,000 - 8,035	62' 2"	623	53 (62) - 180
XCG-10	7,980 - 15,980	105' 0"	1180	50 - 150
CG-13A	8,900 - 19,100	85' 8"	873	80 - 190
CG-10A	12,000 - 32,000	105' 0"	1180	50 (77) - 180

These advanced models were designed with greater survivability and hence reuse than was operationally experienced by the CG-4A. They included the following:
- All-wood construction
- Durable flooring
- On-axis tow plug
- Superior performance specifications (attempted but not always attained)

The design of the CG-4A was good enough to press into wartime service. However, the CG-4A and even its intended successors, the CG-15A and CG-13A, were not laudable engineering examples by modern standards. They were produced with unacceptable performance shortcomings. The baseline model for any modern consideration starts with the last and greatest production cargo glider model—the Laister-Kauffman CG-10A. The operational CG-10A was an impressive feat of engineering. It was high technology for the day and produced by the one vendor. Successfully passing a mature test and acceptance process, at V-J Day the CG-10A was in full-rate production for the upcoming invasion of Japan. The precedents that the CG-10A set for US aircraft include the following:[28]
- First aircraft with rear doors and low cargo floor under a high tail

- First large aircraft to position its landing gear to the sides of the fuselage rather than under the wings
- First aircraft to use quadruple-disk hydraulic brakes
- First aircraft to use thick wing skin as the primary wind bending structure
- Strongest aircraft floor at the time
- First aircraft to carry a 2-1/2 ton truck or a 155mm howitzer
- First aircraft to carry 60 paratroopers
- Largest proven-successful nearly-all-wood aircraft

For the logistics glider, significant wing and cargo geometry improvements are envisioned to the point that the original design might be unrecognizable. The CG-10A attributes to be carried into the logistics glider include the following:
- Rear-loading under a high wing and tail
- Tricycle wheel configuration
- Baseline performance characteristics such as pickup payload capacity, flight speed range, descent rate, and landing accuracy.

Table 4 is a compendium of salient specifications for consideration in iHL baseline modeling.

Table 4. CG-10A Specifications

External Dimensions	70 L x 27 H x 105 W	feet
Cargo Volume	24 L x 6.67 H x 8.5 W	feet
Weight Range	12,000 - 32,000	lbs
Pickup Weight Range	13,350 - 25,000	lbs
Pickup Roll at 13,350 lbs[29]	120	feet
Landing Distance at 25,000 lbs	297 - 650	feet
Maximum Tow Speed	180	mph
Maximum Speed with Flaps	125	mph
Maximum Descent with Flaps[28]	1200	feet per minute
Maximum Lift / Drag	14:1	Ratio
Lift / Drag with Flaps	4:1	ratio
Maximum Flaps	60	degrees
Stall Speed at 13,700 lbs	62	mph
Stall Speed at 13,700 lbs with Flaps	50	mph
Stall Speed at 23,000 lbs	72	mph
Stall Speed at 23,000 lbs with Flaps	62	mph
Stall Speed at 25,000 lbs	77	mph
Stall Speed at 25,000 lbs with Flaps	62	mph
Wing Area	1180	square feet
Wing Loading	21.2	lbs per sq ft
Aspect (span-to-mean-chord)	8.15	ratio
Wing	*Wing Root*	*Wing Tip*
NACA Airfoil	23018	4412
Chords[29]	172 inches	97.5 inches

The snatch pickup attributes to be carried into iHL are:
- Single and multiple cargo glider snatch pickups
- Ground intercept station for austere ground launches
- Complex combination of nylon and steel in the towline
- "Chainsaw" or orbital route of the tow craft's sortie
- Fast, horizontal speed of the tow craft
- Tow craft boom and energy-absorbing winch
- Very short takeoff distance for the cargo glider
- Improved and unimproved runway capability

This historical compendium presents cargo gliders and snatch pickup from an atypical perspective of logistical systems engineering. The historical record is not representative of future iHL implementations or modern logistics glider performance. But it forms an understanding of the basis for the proposed conceptual system, insight into those similar tradeoff decisions yet to be made, and the advantages or consequences from many painful lessons learned.

2.1.2.1 Cargo Glider Epilogue.

Total US production of cargo gliders totaled 14,471.[15] In-theater missions involved 4,058 gliders including reuse. And during the five years of cargo glider snatch operations, 474 in-theater snatch pickups have been identified—with attempts known to have followed at least 12 of 14 missions (details of two more missions remain unknown).

There was no defining moment or decision signaling the end of the cargo glider and its snatch pickup. Wartime production contracts were terminated and development faded. Many in the glider production industry had actually envisioned a bright future in commercial passenger service.[4] The Civil Aeronautics Board would not permit that.[31] In the postwar production slump, the only known commercial application, Winged Cargo Inc., hauled fresh strawberries and tomatoes in war surplus CG-4As from runways in Florida to the northeast. But they didn't last.[11] Rather, bulk transport turned to runway-based powered flight and air assault to helicopters. The Marines developed vertical envelopment in 1947. Helicopters overcame their practical shortfalls first illuminated in the CBI Theater and continue to offer tactical precision in austere transport.

Likewise, the blossoming sea-based supply infrastructure in the Pacific Theater proved unjustified for ensuing expeditionary logistics. The Cold War era established forward bases with a supply chain for invasion forces typically less than 600 miles away. Combining the helicopter with forward land bases essentially masked their individual supply chain disadvantages including centralized depots, high fuel consumption, and short delivery legs. This combination effectively extinguished the expeditionary advantages of cargo gliders and snatch pickup in long-range, austere, distributed precision delivery.

Expeditionary logistics has since changed again as Sea basing supplants forward bases. Ever-increasing roads, airfields, parking lots, stadiums, etc., continue to make LZ selection today less predictable and amphibious landings more so.

2.1.3 **STARS**

The Surface-To-Air-Recovery System (STARS) fielded in the 1960's by All American Engineering was used by the Navy to pick up mailbags and by the National Aeronautics and Space Administration (NASA) for the recovery of missile telemetry recordings. STARS intercepted a disposable balloon to dead-lift the payload from the deck of the ship. Figure 5 is a photograph of a C-130 intercepting the STARS balloon.[5] STARS was phased out in the 1970's with the development of wireless telemetry.

Figure 5. C-130 STARS Intercept

There are center-of-gravity limitations on how much weight can be dead-lift intercepted by an airplane and still maintain its controlled flight. The maximum weight to which the C-130 could be pushed was almost 3,000 lbs depending on its total payload aft of the center axis.[13]

The attributes to be carried into iHL are those of the STARS infrastructure, using a balloon and tether to negate ship motion. Tug modifications include a boom and winch installation and, potentially, a generator to power the winch. Winch technology could initially recycle some surviving Model 80 or Model 120 winches but will more likely require modern engineering to surpass previous specifications.

The Office of Naval Research (ONR) developed the Fulton surface-to-air recovery system in this era, primarily for human and light cargo pickup. Its balloon concept, before the capability retired in 1996, may have matured to be of interest in logistics glider snatch intercept. Overall there were unrealistic modifications to the tug for further consideration in a Sea base application.

2.1.4 Existing Interfaces

Indigenous to the Sea base are exiting air delivery craft (MV-22, CH-53E) with external cargo capability used for expeditionary resupply. For iHL, the intent is to modify some of those Sea base assets for iHL delivery and retrograde action. Additionally, helipad snatch can be performed by land-based C-130's or similar long-range aircraft. This enables joint Air Force, Army, Coast Guard, and Coalition interoperability with the naval Sea base.

The Sea base storage, load, and recovery systems will need to be optimized for iHL capabilities and performance metrics. It is not intended that iHL alter planned ship constructions. Modifications do need to be made, such as automated equipments, replacing welded safety rails with removable chains, and advancing the flight operations capability of most of the ships (lights, markings, procedures, etc.). The low-profile nonskid deck of the *Bob Hope* and *Watson* classes, for example, restricts rotorcraft landings to emergency hover-only but should be favorable to logistics glider takeoffs.

Ashore, USMC Combat Service Support (CSS) echelons have a processing and distribution system for expeditionary replenishment; however, they may not have sufficient equipment, training, and procedures to process resupply if wholly delivered by air or logistics glider in particular. An essential part of logistics glider development will be its design for usability by the CSS to scale from ashore MEB resupply down to direct delivery to the distributed tactical unit.

2.2 Proposed iHL Concept

The iHL concept is first described by a typical flight profile followed by major tradeoff considerations for reuse, occupied flight, and motorized flight.

2.2.1 Flight Profile

The typical flight profile of the loaded logistics glider begins on the helipad of a supply ship. An aerostat lofts the towline attached to the logistics glider on the deck. An orbiting tow craft (tug) in horizontal flight uses an external boom to hook the looped towline. Cable pays off the tug's winch drum as its drum brake carefully tensions the towline until the winch locks. A nylon component in the towline acts as a spring to accelerate the logistics glider off the ship and into tow behind the tow craft. Multiple logistics gliders may be snatched by the one tug per sortie.

Then the tug sorties to a release point. Upon release, the logistics glider descends to an unimproved LZ. The tug returns to the supply ship for a repeat cycle while the cargo is processed on the ground.

When reused, the logistics glider departs the LZ using the WWII-style ground station pickup technique. There is a vertical delivery to the Sea base with the tug sling carrying the logistics glider.

2.2.2 <u>Reuse</u>

The expectation of one-way glider delivery operations during WWII turned out to be unrealistic, and greater emphasis was placed upon survivability and recovery as experience grew. Those lessons learned were incorporated into the CG-10A with advanced delivery survivability features. CCAAF certified CG-10A snatch to payload capacities significantly above any consideration of a simple empty retrograde. This implied the recognition of glider snatch for operations other than retrograde after its initial delivery. Perhaps this meant in-JOA delivery routes or snatch out of unimproved forward operating bases (the beginnings of vertical tactics).

For STOM, a typical MEB resupply scenario is expected to have dozens if not hundreds of vehicles (tugs and logistics gliders) in circulation on a daily basis. System elements or handling equipments that are disposable—such as logistics gliders, balloons, rocket assists, and even tug fuel—have to be supplied into the Sea base taking up limited deck space or cargo capacity. For snatch pickup, the tug burns fuel in the more efficient horizontal flight mode and the towline can be recycled.

The iHL developmental intent is to demonstrate a reusable prototype, from the practical standpoint of the costs of prototype airframes, the likelihood of continued technology insertion, and the wisdom from WWII. If at a future decision point the full rate production economics and Sea base processing environment justifies a disposable technology for a logistics glider, then the proposed developmental effort in this report lays the fundamental groundwork to support such an approach.

2.2.3 <u>Unpiloted not Unoccupied</u>

The fundamental iHL concept does not require nor exclude glider pilots. Any occupied flight operation demands acceptable reliability and realistic safety procedures beyond the concept demonstration efforts in this report. Helipad snatch is expected to be near the performance envelope on weight and launch stresses, and there is a lack of safe alternatives once accelerating across the helipad. As a lesson learned, attempts in the 1940's to combine glider snatch with commercial passenger service was something the Civil Aeronautics Board consistently ruled against.[31] A threshold capability could involve piloting of some kind. But iHL is driven by the efficient transport of cargo without non-consumable items delivered to the LZ. The objective of bulk cargo delivery is not to require onboard crew or external remote control, but many factors will have to be weighed prior to operations.

However, there are other, less challenging launch environments in which a large-volume air vehicle can operate and still be compatible with at-sea helipad launch. Once in the JOA there are undoubtedly many other applications beyond ship-launched delivery. These are only limited by the helipad launch requirement. With such a large volume transporter, tactics from the WWII successes in glider snatch, or more modern rotorcraft operations will be applied. Given sufficient safety margins, these concepts are categorized here to allow options for human occupants upon takeoff or require an optional pilot for landing. Certifications notwithstanding, the operational vehicle should be physically able to support these three possibilities:

- Unoccupied operation (required) with
  - Autonomous control or
  - Remote control or
  - Hybrid autonomous and remote control
- Piloted (desired)
- Passengers onboard (desired; procedures may dictate a pilot)

Concept demonstration vehicles are likely to be piloted until the incorporation of any technology for autonomous, remotely controlled, or hybrid navigation and flight control systems. Crew is always expected aboard the tow craft.

2.2.4 **Motors**

Logistics glider takeoff, flight, or taxiing may be enhanced with rocket assist or integral engines, provided there are acceptable tradeoffs for unit cost, system performance, and Sea base support footprint. Helipad launch will always be the primary performance driver, and every bit of logistics glider surface will have achieving flight in a restricted distance as a priority. Otherwise, secondary priorities can be addressed in vehicle design, such as in-flight power sources, integral propulsion to the delivery point, and over-weight launches.

2.2.4.1 Internal Power.

There is very little power required by a logistics glider sitting or accelerating on a helipad. Ideally a ground vehicle-compatible battery should be used. But once there is sufficient airflow to allow control surfaces to operate, more power may be required. This is a consideration to add to engineless, towed glider situations.

Essentially a windmill, a ram air turbine or RAT has a small impeller that feathers in the slipstream to generate power for onboard systems. There will be many priorities to resolve when attaching a propeller on the airframe, such as drag at liftoff and damage from towline breaks.

If, like the CG-10A, the rear wheels do not retract, then their hubcaps could be designed to feather the wheels within the upper range of towed flight speeds for some power generation.

2.2.4.2 Small Engines, High Performance.

Reducing the tug's delivery involvement to just the helipad launch and consequent acceleration of the logistics glider to its cruise speed, a small combustion engine is sufficient to then maintain a logistics glider's momentum onward to the landing zone. This concept can be a reliable delivery system for performance improvement beyond multiple snatch pickup sorties, in that tug resources are tied up in resupply operations in very short orbit cycles.

The planning of the logistics glider's free flight after tow is a critical design decision. This design approach is in contrast to the volplane scenarios shown in Table 14, "Towed Delivery Scenarios."

2.2.4.3 JATO.

Like everything else that flies, a JATO rocket motor can be strapped on a logistics glider if structurally certified for such. Primitive attempts on WWII cargo gliders[10] did not lead to operational implementation. Relying upon a non-renewable launch source or energy inefficient means for many routine operations is not a realistic use of the Sea base footprint. However, there may be limited applications, such as overweight or high-reliability launches (e.g., occupied flights) for which the extra boost at launch may be recommended.

The timing and amount of additional thrust during both start and cutoff is critical to preventing undue stresses upon the towline. The large variance in vehicle weight due to payload will also have to be taken into consideration when applying rocket thrust. Structural considerations will be minor, as the acceleration would be the same as becoming airborne across a helipad by glider snatch.

2.3 Notional Vehicles

A range of logistics glider concepts frame the design trade space for future efforts. This starts with PHST considerations for the vehicle aboard the Sea base and then engineers the necessary aviation physics. This way the iHL system fits into the Sea base as a performance connector between surface and ground logistics chains.

The entire logistics glider vehicle's surface is optimized to achieve lifting flight in as little takeoff roll distance possible, yet still supports towed flight at high speeds. Any logistics glider design will require the following:
- An airframe with high bending strength and stiffness as well as torsion strength and rigidity
- Achieving flight rotation speed quickly—short takeoff and landing
- Good climb characteristics with high lift and low drag at low speed
- Full control of the aircraft throughout the entire speed range
- High cruise speed with low drag

Significant modification to one or all of these concept designs will probably occur. Each design is presented to facilitate an understanding of specific capabilities, some of which may be transferable to the other designs or a new combination of logistics glider(s). For instance, propulsion may be added to any concept, but is just described on the final vehicle demonstrating an application of additional launch forces beyond helipad snatch equipments.

The USAF designation XG (experimental glider) supersedes the USAAF designation of XCG (experimental cargo glider). Considering the planned CG (X) surface cruiser currently under development, this report uses the XG designation and resumes numbering after cargo glider development ended with the XCG-20. Not coincidentally the one-TEU volume logistics glider is designated XG-21, the two-TEU volume XG-22, etc.

The XG-21 is the smallest logistics glider model and lowest risk toward threshold demonstration. However, it has cargo and retrograde shortcomings that may prevent the sole resupply of a MEB. The more robust XG-22 concept vehicle is the baseline configuration used to describe logistics glider operational concepts in this report. It is envisioned as the demonstration basis to transition. Implementation approaches are in the Tactical Lifecycle section to follow. The XG-22 is also the recommended basis to develop further concepts, such as those described afterwards: XG-23 and XG-24 appendage to the XG-22 and the XG-3x series that straps onto its payload.

The artistic concept pictures and any design implications to follow are not intended to convey any structural engineering decisions. These renderings depict a body-on-frame construction approach, but a monocoque (e.g., unibody) approach or a hybrid construction of partially isolated sub frames should also be considered within the footprint restrictions. Structural and materials design considerations need to include the Sea base repair and austere landing and launch environments. The intent is to show options for processing, a low assembly count, and the general "look and feel" for those not familiar with glider design.

2.3.1 ISO-Based Assembly

The largest standard shape that a cargo ship processes is the ISO standard TEU container. Any larger or nonstandard profile creates additional processing complexity negating its advantages to the Sea base. The logistics glider vehicle itself should be packaged to meet ISO handling standards to efficiently integrate with supply ship operation.

The PHST of the logistics glider exterior drives the utility of the system: There simply is no extra parking space at the Sea base for dozens if not hundreds of these vehicles. The objective is to fit the airframe into multiples of TEU packages for storing each as ships cargo and have standardized lift points for handling. Given standard TEU handling points and appropriate mounting brackets and supports, they can be brought aboard, stored, moved, and even stacked on top of each other.

Likewise, the payload capacity of the logistics glider is sized to transport and access the specific Sea base deliverables to the warfighter: vehicles, artillery, bladders of liquids, and multiples of JMIC pallets.

Table 5 summarizes the concept vehicles described here with gross estimates of their maximum lift capacity. The "Footprint" column is the TEU count aboard the Sea base that the logistics glider requires in its stored configuration. The "Payload" column is an estimated capacity model of 60% from Table 14, "Glider Payload Ranges," and within the range calculated in Table 13, "Airfoil Lift Capacity" of the lifting surface area that can be fit into the stored footprint. This is representative until such time as supporting engineering actually calculates these weights and shapes to be structurally viable and physically flyable. The "JMIC quantity" column uses each cargo hold's respective dimensions, and the "Max Capacity" column provides examples of the designed maximum shape of its payload.

Table 5. Logistics Glider Size and Payloads

Logistics Glider	Footprint (TEU)	Payload (lbs)	JMIC qty	Max Capacity
XG-21	1	6,800	5	JMIC
XG-22	2	13,500	16	HMMWV (LW155)
XG-23	3	24,000	16	HMMWV
XG-24	4	22,000	32	HMMWV
XG-31	1	6,800	6	JMIC
XG-32	2	13,500	16	HMMWV / LW155
XG-33	3	24,000	16	HMMWV / LW155
Amphibious	-	32,000	42	LAV / MTVR

The maximum capacity of the XG-22 includes a custom modification to house an LW155 howitzer as described after the descriptions of the logistics glider variants.

For a modeling placeholder, the following additional assumptions are made in the table. The three bi-wing models (the XG-23, XG-24, and XG-33 logistics gliders) benefit from the second wing's lift and account for it and the associated supporting structure by subtracting 1,500 lbs from the gross vehicle weight available for payload. The XG-24 payload model estimates the additional fuselage (behind the center of gravity) by subtracting 3,000 lbs off the gross vehicle weight before calculating the payload percentage. Any increase in payload capacity of the XG-3x series by using a standard container in lieu of a fuselage requires a structural design analysis.

2.3.1.1 XG-21 Logistics Glider.

This logistics glider comes packaged with all its components for assembly and helipad launch aboard the Sea base within a single ISO standard TEU container. These components are listed in Table 6. The fuselage includes the nose and rear, unlike the other ISO-based concepts to follow. Within the stored fuselage are its retractable wheels, the horizontal stabilizer and the fuselage bottom.

The three wing sections each have specific shapes but take up identical storage volume as listed in the table. The middle wing attaches to the top of the fuselage and the left and right wings are added for a wingspan of 55 feet. The single tail boom and vertical stabilizer component lays sideways in storage and assembles onto the rear of the fuselage for a total length of 38 feet. The horizontal stabilizer attaches to the top of the vertical stabilizer, out of reach from casual contact.

Table 6. XG-21 Logistics Glider Assemblies

Component	Qty	Outside Dimensions	Lift Area (ea)
Fuselage – stored	1	19' 4" L x 7' 8" W x 3' 10" H	-
Wing	3	19' 4" L x 7' 6" W x 10" H	139 sq ft
Tail	1	19' 4" L x 10" W x 7' 6" H	-
Horizontal stabilizer	1	13' 0" L x 7' 0" W x 8" H	91 sq ft
Bottom	1	13' 0" L x 7' 0" W x 6" H	-

The XG-21 has three wheels: a turning wheel centered in the front nose and two attached to the fuselage near the rear. All are retractable both in flight and during packaged storage. The nose cone in Figure 6 is opened upwards and its wheel is extended.

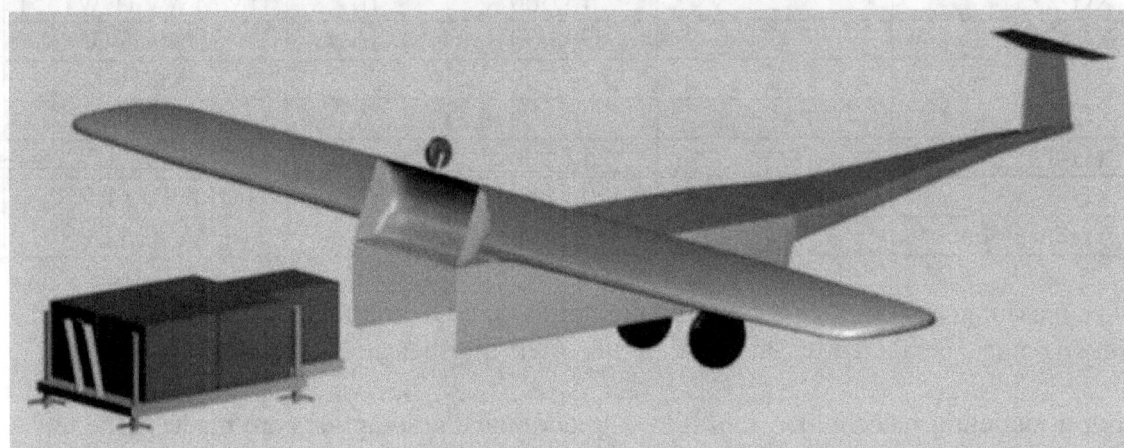

Figure 6. XG-21 Cargo Access

The internal dimensions are exactly sufficient to store five JMIC containers, not all of which can fly at the maximum specified weight. This cargo bay is too small for a person to work inside. Rather, its payload is accessed as conceptually envisioned in the figure by separating the bottom from the fuselage. The bottom is disconnected and stands free on jack stands for efficient loading and access. The nose is hinged up so the fuselage is pushed or towed backwards away from the bottom.

There is very little design leeway on the thickness of the fuselage sides. There is approximately 4.5 inches between the inside width of an ISO standard TEU container and the outside of two JMIC units. This has to accommodate two glider sides along with the sliding leeway of the fuselage in and out of its container as well as the sliding leeway of multiple JMIC containers in and out of the XG-21 cargo bay.

2.3.1.2 Baseline XG-22 Logistics Glider.

This logistics glider concept consumes a two TEU footprint aboard the Sea base. It is the baseline default design used in this report. Its six wing sections listed in Table 7 fill one ISO standard TEU container and are removed for assembly into a mono wing atop the glider fuselage. They are attached along the sides of the top of the fuselage body. Including the body's 8-foot width, the total wingspan is about 60 feet.

Table 7. XG-22 Logistics Glider Assemblies

Component	Qty	Outside Dimensions	Lift Area (ea)
Fuselage	1	20′ 0″ L x 8′ 0″ W x 8′ 0″ H	120 sq ft
Wing, main	4	19′ 4″ L x 7′ 6″ W x 1′ 8″ H	145 sq ft
Wing, tip with ailerons	2	9′ 8″ L x 7′ 0″ W x 1′ 4″ H	67 sq ft
Nose	1	6′ 0″ L x 7′ 6″ W x 7′ 6″ H	-
Tail	1	16′ 0″ L x 7′ 6″ W x 1′ 4″ H	120 sq ft
Bottom	1	20′ 0″ L x 8′ 0″ W x 1′ 6″ H	-

The remaining, non-wing vehicle components listed in the table come prepackaged inside the cutaway fuselage body visualized in Figure 7. The fuselage is the second TEU volume. It is the aerodynamic equivalent of an 8x8.5x20 foot container with lifting body and wing connections integrated into the top. Mounting hardware and supports may be required for handling and for stacking when stored. The nose cone (green), rear doors (blue), bottom (purple), and tail (red) components are stored inside the TEU.

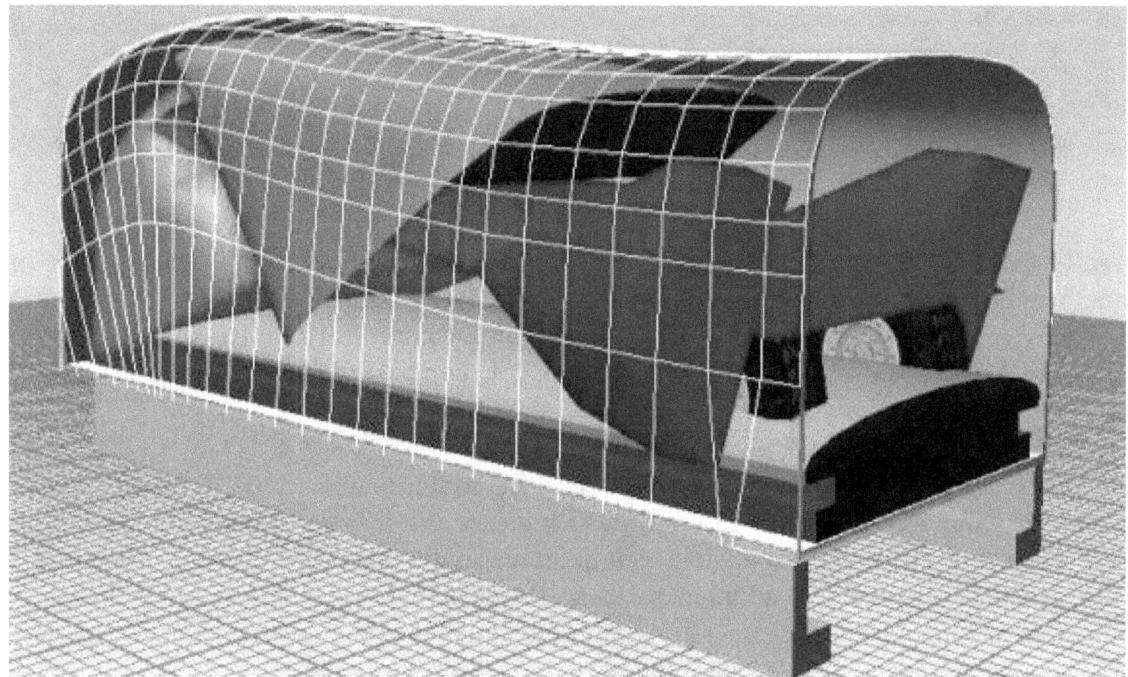

Figure 7. XG-22 Packaged Fuselage Rear Quarter View

Figure 8 shows the fully assembled XG-22 logistics glider. The rear doors, nose cone, and tail sections are assembled into operating positions. The frame is jacked up and the bottom slides out and is flipped over for reinsertion as the airframe bottom (note bottom of Figure 7). A replaceable nose skid (not shown) may be required. At least two wheels are under the body near the rear of the bottom. Likely there would be a single wheel centered in the front, only canted as far back as practicable. This way its footprint maximizes the width of the helipad available during snatch takeoff roll.

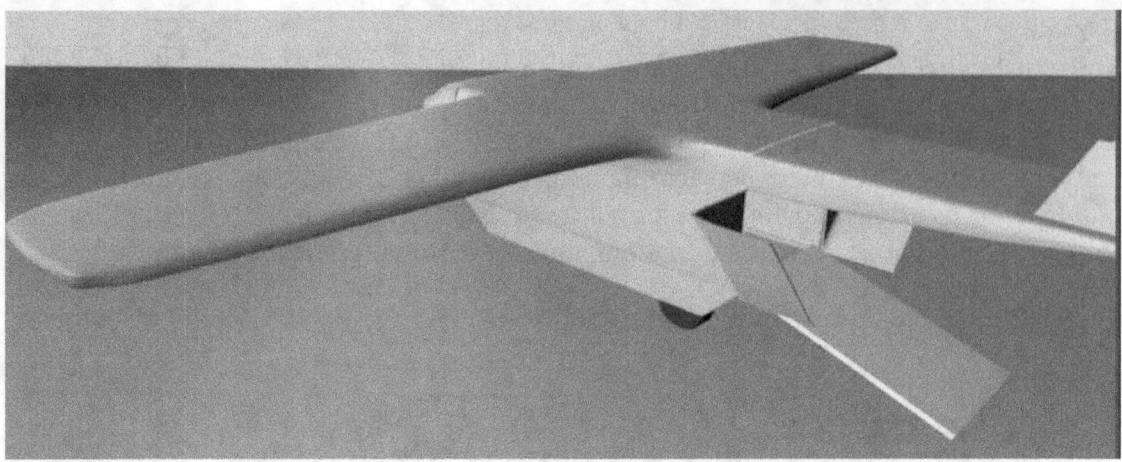

Figure 8. XG-22 Baseline Logistics Glider Rear Quarter View

While the XG-22 stores in two TEU volumes aboard the Sea base, some content of another TEU container supplies its first payload. The XG-22 is intended to carry a HMMWV, or combinations of JMIC, pallets, liquid cargo, and other shapes such as Class VIII repair items that fit within a standard ISO container.

2.3.1.3 XG-23 Logistics Glider.

This logistics glider concept consumes a three TEU footprint aboard the Sea base. It adds a second wing above the XG-22 concept to increase the lifting surface and consequent payload weight with minimal vehicle weight increase. It consists of six wing components and structural members in addition to those listed in Table 7. The bi-wing design is intended to support launch from a helipad. This logistics glider's wings come inside two TEU containers while its fuselage is the third TEU footprint in this design.

2.3.1.4 XG-24 Logistics Glider.

This logistics glider concept consumes a four TEU footprint aboard the Sea base. It adds one fuselage and six wing components to those listed in Table 7, doubling the TEU footprint. Essentially a second XG-22 or XG-23 fuselage is appended to the first for a limited-capacity but flight-worthy vehicle. Due to the longer combined wheelbase there is less distance to accelerate across a helipad. Potentially too long for helipad launch, this concept is intended for ground- or runway-based takeoff. Significantly less payload weight is supported since the extension counts against GVW and additionally pulls the center of gravity aft. The XG-24 concept supports:

- Lighter payloads emphasizing volume over weight, such as passengers or oddly shaped specialty items.
- For retrograde return to the Sea base, this doubles the logistics glider count per sortie so as to halve the number of VERTREP delivery missions.
- Delivery by air into the JOA from an advance base by half as many delivery sorties as other designs. This reduces the transport footprint aboard the Sea base.

There are several alternatives with XG-24 wing implementation for two flyable craft into one. A bi-wing approach is described in the XG-23. A grocery cart analogy is chosen for the XG-24: The nose of the second XG-22 is appended to the rear of the first's fuselage. The first's tail and rear components are placed inside. Wings remain in place and operable, but control and lift are altered and not as effective. The implementation of any XG-24 concept requires significant structural and aeronautical design verification.

2.3.2 Strap-on Logistics Glider

This advanced logistics glider concept is disposable as an objective, or contains mostly low cost throwaway components, depending on the economics of iHL. Proposed are two mono-wing concepts and one bi-wing concept as multiples of the TEU when stored. Engineered structural cardboard technology, for example, has made significant progress in the past decade, while more traditional low-cost flight structures such as plastics, laminates, or particle woods are more likely candidates.

Inexpensive or inflatable flight components attach to the payload of a standard container or containers that have been prepared for air delivery. A TEU container would not return to the Sea base. Wheels will need to be added for at least launch from a helipad, and the concepts represented here depict HMMWV-compatible tires for movement in the field.

Payloads are restricted to either fitting inside a TEU container (e.g., JMIC and HMMWV) or suspended from the strap-on airfoil in lieu of the TEU container and the rear cowling (e.g., LW155 howitzer). With the reduced vehicle structure, it is envisioned that a greater percentage of GVW can be payload compared to the XG-2x series. This greater density makes for fewer vehicles launched daily in resupply. However, an empty TEU container weighs 3,500 lbs and cannot be considered useable load.

Without the aerodynamic shaping of the fuselage and retractable landing gear, the aerodynamic drag of the strap-on logistics gliders will be higher than the XG-2x series, negatively impacting tow speeds and glide ratios. Steeper glide slopes translate into further towing at lower redline speeds, increasing the required tug asset count when comparing similarly sized payloads.

2.3.2.1 XG-31 Logistics Glider.

This strap-on logistics glider concept consumes at most one single ISO standard TEU container aboard the Sea base. Ideally the XG-31 is designed to store two vehicles in a single TEU container. It has components similar to the XG-21 concept except that the fuselage has minimal stored height without sides, and the rear wheels do not retract into the vehicle. The XG-31 takes a different structural and assembly approach in that the fuselage and wing become one lifting body top, attaching to frames in the nose and tail. The bottom is then an impact skid with wheels.

Under its lifting airframe, the XG-31 suspends a payload of three pairs of JMIC containers. Their exterior sides are exposed since the XG-31 has none. By removing the top of the body, JMIC can be accessed in the following ways:
- In situ through the container top like so many bins
- Emplaced or removed by forklift

Like the XG-21, the XG-31 useable payload capacity limits each JMIC container to an average one third of the maximum JMIC weight specification. The six containers themselves as payload remove an estimated 1,800 lbs from useable payload.

2.3.2.2 XG-32 Logistics Glider.

This strap-on logistics glider concept stores in a two TEU footprint aboard the Sea base. A six-section wing is assembled out of one ISO standard TEU container. The lifting body, tail boom, tires, and inflatable nose and rear cowlings come out of another.

Its typical payload is one ISO standard TEU container. For disposable or the first delivery flight, this is a third TEU volume stored aboard the Sea base. The tires, inflated nose and rear cowling can be disposed after use. The wing, lifting body, and tail boom may be recycled.

2.3.2.3 XG-33 Logistics Glider.

This strap-on logistics glider concept consumes a three TEU footprint aboard the Sea base. It adds a second wing above the XG-32 strap-on concept to increase the lifting surface and consequent payload with minimal weight increase. Including structural members, the bi-wing concept is intended to support launch from a helipad.

Figure 9 displays this bi-wing configuration to demonstrate how multiple wings might increase wing lifting area within the helipad's area constraint. The implementation of this concept requires significant structural and aeronautical design verification. These two notional views display merely the conceptual assembly and strap-on shapes for furthering discussion.

The XG-33 strap-on components come in three ISO standard TEU containers. Its payload consumes a fourth TEU footprint for its initial delivery or for every disposable delivery. The container is shown in blue in the figure and represents any ISO standard TEU container.

Figure 9. XG-33 Strap-on Glider Views

2.3.3 <u>Amphibious Logistics Glider</u>

The seaplane is the future of naval platforms in the development of distributed and autonomous operations. The amphibious logistics glider is based upon the 1950 Navy R3Y-1 Convair Tradewind concept only without flight engines. This non-ISO volume seaplane glider is surface-snatched carrying a Medium Tactical Vehicle Replacement (MTVR) or Light Armored Vehicles (LAV) and air-towed to riverine or unimproved ground delivery. It is seaworthy for at-sea anchoring, towing, snatch pickup, and amphibious delivery. It can be loaded by any of the following methods for flight-worthy snatch pickup.

- Preloaded for runway tow, surface tow, or cargo transport into the littoral
- Loaded on a supply ship and placed in the littoral waters by crane
- Loaded on an MLP and rolled into the littoral waters
- Filled afloat, either beside a liquid supply source or towed such as by an oiler

Rather than the helipad snatch technique, a disposable balloon is intercepted by the tug for a surface snatch nearby the Sea base. Its retrograde delivery is by landing on the surface near the Sea base rather than be air-delivered to the Sea base helipad.

The amphibious logistics glider maneuvers in the water via a small engine with either an air fan or a directional pusher propeller in the water. Figure 10 opts for the propeller under the water line in the rear. Design options include the wings folding for on-deck storage and processing, JATO launch, and motor glider flight and taxi capability.

Figure 10. Amphibious Logistics Glider Front Quarter View

Its payload capacity supports the size and weight of one MTVR at 31,069 lbs, one LAV at 28,200 lbs, or as an objective requirement, potentially one Stryker at 19 or higher tons. Without helipad launch there is no wheel print or wingspan constraint. So a larger mono-wing and non-ISO standard fuselage is designed around the MTRV payload capacity shown in Table 13, "Airfoil Lift Capacity." Amphibious logistics glider wingspan is approximately 165 ft as derived from the requisite lifting surface area.

2.3.4 Custom Applications

There are numerous concepts that entail modifying a basic logistics glider configuration. Several are described here. These change its general purpose cargo nature into a specialized capability.

2.3.4.1 LW155 Howitzer.

The M777A1 Lightweight 155mm howitzer is currently fielding to the Marines and Army. A fire control computer is being retrofitted to an earlier version, raising its towed weight from 9,200 lbs to approximately 9,800 lbs. It has stowed, towed, firing configurations, and potentially a sling-carried configuration. The sling load transport configuration is similar to if not identical to a level firing position: Its rear spades are extended in a spread-eagle configuration. The towed position has the rear spades folded in and over the rear so its widest dimension is 8.5 ft. This is across both the wheels and across the folded spade's elbows.

The LW155's stowed configuration includes a third wheel weighing as much as 60 lbs. It is easily removed and would not be delivered to nor used by combatants in the field.

The strap-on logistics glider concept can suspend the LW155 from the top airfoil structure and eliminate any bottom, side, and rear glider components. There will be a reduced maximum tow speed due to the uneven airflow, and the wheels required at launch and landing will need assessment. The minimal-component XG-32 depiction in Figure 11 uses the LW155's wheels.

Figure 11. XG-32 Strap-on Logistics Glider with LW155 Howitzer

Modifying the XG-22 logistics glider fuselage to contain the LW155 is desirable for high tow speeds. This will require careful measurement and a custom logistics glider bottom of high design complexity. The spades may have to be removed during transport within the XG-22 and reassembled upon delivery. It is unclear if also removing the wheels will fit it into an ISO standard cargo volume. Retrograde transport may have to consider fitting an LW155 that is dirty, damaged, or otherwise out of specified shape or center of gravity.

A custom, wider XG-22 bottom and raising the fuselage higher can accommodate the LW155 in its towed configuration. It is wider for a depth of at least 30.75 in. for the LW155's trails. The bottom is assembled at the Sea base and would not normally be removed in the field. The LW155 is rolled into the logistics glider rear first in a stowed configuration sans the third wheel. The barrel sticks out modified rear doors. The LW155's weight can alternatively be suspended from structural points in the ceiling for a sling load carry rather than the bottom and fuselage frame.

2.3.4.2 Fueling Station.

Multiple large bladders, pumps, and hoses can run from the logistics glider to refuel multiple vehicles parked safely around a refueling operation. This also applies to water, other wet payloads, or specialized transport missions. Figure 12 is a photograph of a CG-4A used as a water tanker in Sicily or North Africa during WWII.

CG-4A USAAF photo, Roland Fetters collection from Charles L. Day, photo contrast is modified

Figure 12. WWII Water Tanker

2.3.4.3 Medical Facility.

Figure 13 is a photograph of one of the WWII European theater glider ambulances used at Remagen.[32] They transported 700 to 800 American casualties to hospitals behind the front lines. Twenty-five casualties were snatched including enemy wounded[3] just prior to Operation Varsity.[24] Six similar ambulances were prepared for CBI theater casualty evacuations[4] and at least four were used.[18]

Figure 13. WWII Ambulance

A logistics glider converted into a mobile medical facility allows the doctor to make house calls. Used as an ambulance, evacuation of the entire facility with patients can occur quickly for delivery to a larger medical facility potentially at the Sea base or, more likely, elsewhere. Any evacuation can occur with minimal, high-speed tug exposure.

2.3.4.4 Sensor Platform.

Logistics gliders can be equipped with nearly any type of sensor for battlefield surveillance. While an individual logistics glider may not spend enough time in flight over any particular area to justify the added complexity, cost, or risk, a staggered series of delivery flights can together provide valuable information. This surveillance tactic is especially relevant to the tug entering a contested zone behind the logistics glider for pickup or other missions. Additionally the logistics glider in this mode of operation can disperse longer-lived surveillance platforms on demand.

A logistics glider can perform a one-way standoff delivery as an unmanned Chemical/Biological/Radiological hazard-sensing suite and communications platform into a suspect area. It is a good choice as a wash station for contamination cleanup.

2.3.4.5 Aerial Reload.

The Air Force Research Laboratory has an advanced concept for the reload of munitions in flight.[33] Using munitions supplied by the Sea base, the naval equivalent can be a similarly outfitted logistics glider snatched and then towed to reload combat aircraft. The logistics glider would not be released by the vertical tow craft, but rather retrograde to the Sea base for reload once logistics glider weight drops within range for a vertical carry.

2.3.4.6 Towed Gunship.

A towed glider can be turned into an autonomous close air support gunship. Several gunship gliders may be towed by a single tug for economical air-to-ground fires. The bottom section is specifically designed as a gun mount while the remaining cargo capacity holds ammunition and automated feeding equipments for unmanned, remotely controlled operation. Without an onboard crew some limitations due to safety, noise, and vibration can be eased.

2.4 Marine Resupply Ashore

The proposed iHL system is one option toward efficient resupply of a Year 2015 Marine Expeditionary Brigade ashore by the Sea base. Other proven methods can perform resupply given favorable conditions; iHL is highly desirable both alone and in conjunction with those other methods in trading off the decisions for the best choices for sea-based resupply performance, scale, distributed operations, and total life cycle costs.

The two metrics to measure sea-based resupply performance are throughput and synchronization.[2] Synchronization with the warfighter for delivery is directly impacted by Sea base operational availability. iHL throughput of materiel quantity is such that the Sea base can resupply an SBME MEB ashore.

2.4.1 Sea Base Operational Availability

An all-aerial delivery approach cuts down many of the following cargo transfer links in the existing supply chain by using air transport directly off the supply ship to the distributed warfighter.
- Ship-to-ship
- Repeated strike downs and strike ups aboard ship
- Surface-to-ground
- Shore depots
- Ground distribution

Minimizing or eliminating these cargo transfer links has a significant impact on supply chain synchronization, specifically the timely delivery by the Sea base of requested supplies. iHL reduces the influence of high sea states as a root cause for missed delivery and conservative reliance upon resupply. Sea state limitations at each transfer point within the supply chain can preclude the timely delivery of expeditionary logistics to the warfighter. The objective goal of most surface cargo transfers is operating in sea state 3.

The Landing Craft Air Cushion (LCAC) is a surface delivery vehicle that cannot operate above sea state 3. VERTREP operations are possible in sea state 4, and it is expected that iHL operations be performed in at least sea state 4 with an objective of sea state 5. Sea state 5 capability yields over 99% availability in the littoral and roughly 80% availability in the open ocean.

Figure 14 is a JOA cross section showing the concept of operating zones and their probabilities of operational availability for expeditionary logistics. The block step lines represent the globally averaged probability density function of sea state (sea state 4 is blue and sea state 3 is green) and two notional probability density functions: security (red) and delivery timeliness (black).

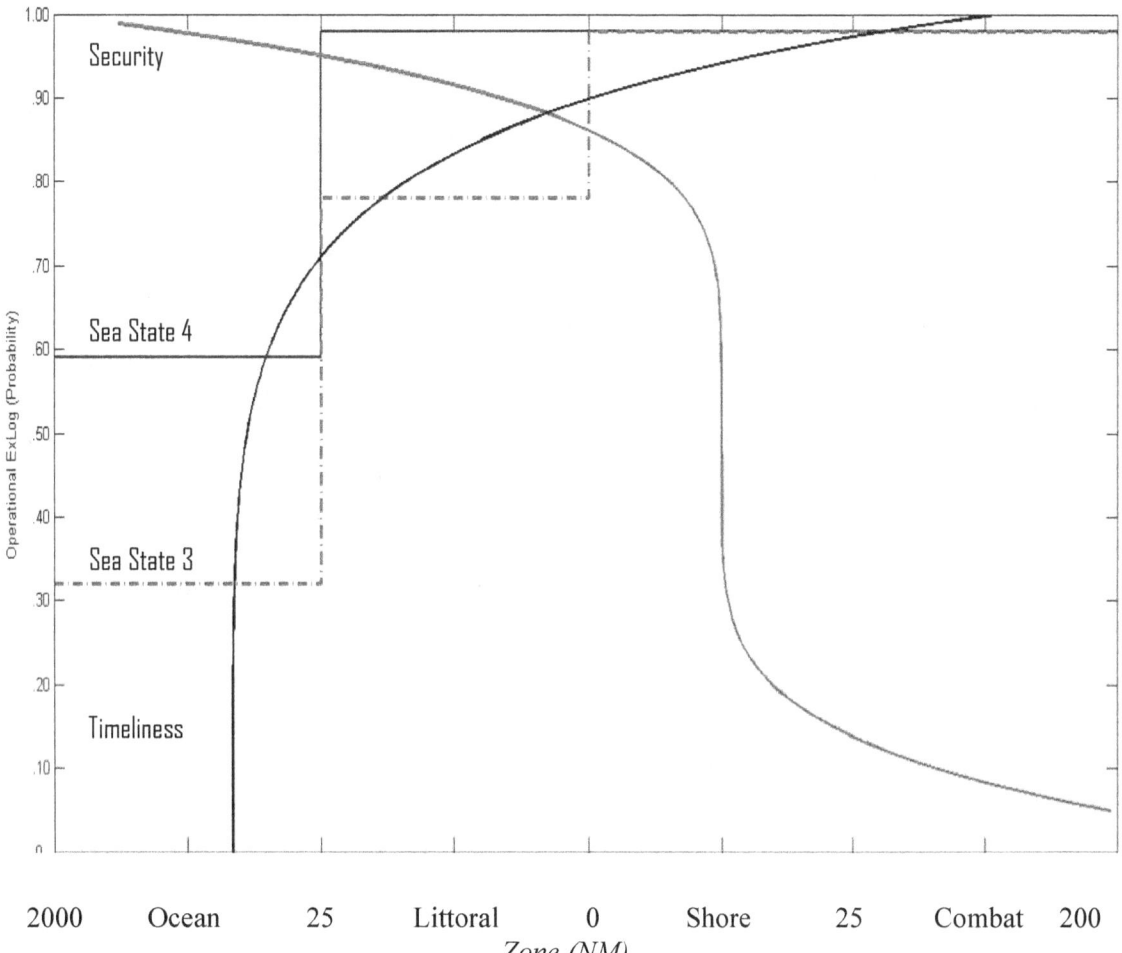

Figure 14. Probabilities of Expeditionary Logistics

The sweet spot under all the curves for operational ability is dominated by the littoral zone, as expected for the chosen location for Sea base maneuver. Sea state 3 or less occurs 78% of the time, indicating any Sea base supply link unable to perform above sea state 3 is unavailable 22% of the time. Sea state 4 or less occurs 98% of the time. The notion is that security is best away from the shoreline but drops significantly once expeditionary logistics move onto land and closer to combat operations. Timeliness in logistics delivery is the opposite—very effective closer to combat operations but dropping significantly with greater distance.

For beachhead logistics, security is highly variable and the chart is intentionally vague to reflect this. Shore depots will remain a requisite buffer within the Sea base distribution chain as long as there is uncertainty in littoral sea state conditions during resupply operations and hence uncertainty in the operational availability of the Sea base. By iHL operating at higher sea states, Sea base operational availability increases, thus decreasing the following.

- The uncertainty in timely delivery (This has implications in the quantities carried by the warfighter and the waste in oversupply.)
- The size of shore depots and their ancillary requirements

Minimizing or eliminating surface and ground cargo transfer links significantly improves not only operational availability, but also supply throughput performance, which is very desirable for the limited resources of the Sea base.

2.4.2 <u>SBME MEB Throughput</u>

If only a single cargo delivery vehicle model is used for all-air sustainment on days 2 through 14 after employment, Table 8 estimates the number of air launches necessary to resupply the SBME MEB. The projected daily sustainment for the sea-based maneuver element of a Year 2015 SBME MEB of 4,989 Marines ashore[34] is 367 short tons.[35] The "Typical Payload" column is based upon an averaged payload of 3.75 short tons per an MV-22 resupply sortie.[36] Being approximately 50% of the MV-22's maximum rated capacity, this is used as a payload density model for estimating other models' options for typical daily launches in their all-air resupply scenarios. The "Total Daily Launches" column is evenly divided among six dedicated helipads in an eight-hour "day."

Table 8. All-Aerial MEB Sustainment

Air Delivery Option	Typical Payload (lbs)	Total Daily Launches	6-Helipad Cycle (minutes)	Maximum Capacity (Class VII)
MV-22	7,500	98	29	HMMWV/LW155
CH-53E	16,000	46	63	LAV/MTVR
C-130T	31,900	23	125	LAV
XG-21 Logistics Glider	3,400	215	13	JMIC
XG-22 Logistics Glider	6,800	108	27	HMMWV(LW155)
XG-23 Logistics Glider	12,100	60	48	HMMWV(LW155)
Amphibious Log Glider	16,000	46	63	LAV/MTVR

The C-130T payload is calculated from Table 15, "Glider Payload Ranges," and is included for comparison purposes only since the C-130 is not part of the Sea base nor can it operate from a helipad. The LW155 requires a custom bottom to the XG-22 and XG-23 logistics glider variants, while the XG-32 and XG-33 are designed to support it. The launch counts and helipad cycle times for the MV-22 and CH-53E air delivery options in the table are presumed unsustainable, which is but one good reason why an all-air resupply concept is not currently viable.

The launch cycle time for each of six supply ship helipads tolerates little variance per cycle. However, if iHL is developed as proposed in this report, there is no technical reason for the following performance-enhancing techniques.

- Multiple snatch sorties to reduce the number of tug assets involved in iHL operations
- iHL performed in conjunction with other (air and surface) delivery options
- Extending iHL across multiple shifts or a 24-hour day especially if designed with automated assistance

The loading and launching cycle times for all of the logistics glider variants are realistic for all-aerial resupply given that automated loading and control is designed into the iHL system. The exception may be an all-XG-21 resupply concept. Although the ALDS concept proposes a 2-minute cycle time,37 the XG-21 cycle time of 13 minutes may be unrealistic for aerostat cycling and logistics glider launch preparation. There are many ways to approach this shortfall and still make XG-21 delivery possible.

- The XG-21 is presented as the minimal capability in iHL implementation. Its payload volume is too small for all resupply shapes. So an initial approach may have the XG-21 supplement rather than wholly provide all-air resupply to an SBME MEB. This means all-air resupply to smaller echelons and distributed units.
- Disposable balloons are not excluded from iHL, and substantial automation is conceivable. It is a price tradeoff on developing that level of performance.
- Some performance criteria can be eased, such as lengthening launch operations beyond an eight-hour day or utilizing additional helipads.
- Any increase in the payload models increases XG-21 performance.
  - o Maximizing the loading model to full capacity places it in the same cycle time as the 50% loaded XG-22.
  - o Increasing the XG-21 usable payload percentage above 60%

As a scenario the table does not include retrograde processing, and so reflects either the first day of resupply using reusable logistics gliders or the typical day using disposable logistics gliders. It is a simplifying assumption that, if needed, retrograde processing occurs during the other shifts each day. Excepting the amphibious approach, the helipad is the bottleneck to all-air resupply of the MEB using iHL.

There are as many as 12 ships in the Sea base for which helipad snatch capability could be mixed in with other operations. A longer day allows for other concurrent operations and doubles or triples the load and launch cycle time for all the logistics glider variations.

Those estimations carried into the typical payload model for the logistics glider variants are based upon the high end of the possible ranges in Table 13, "Airfoil Lift Capacity." The 50% density model could be considered a reasonable delivery model with allowances for standardized packaging weights, spillage, and loss. However, it is very conservative as it is highly desirable to skew this average upwards toward significantly higher payload densities. A 300-lb average increase in typical payload increased the cycle time by only a half minute yet reduced by one tug asset daily in the next table. This shows iHL operates best with standardized containers, automated loading, performance optimization, and as a standardized vehicle for its PHST.

All the logistics glider variants have the ability of intermixing with each other or with other delivery systems. All variants additionally have the ability of double or potentially even triple towed sorties to reduce the sortie count ashore to one-half or potentially one-third of the total daily launch counts of Table 8. The following explores this further and computes any required retrograde assets.

A conventional VERTREP ashore sortie 50 NM away is presumed to simply take 2 hours 40 minutes in this example allowing three complete round trips. This requires at least 33 VERTREP vehicles to resupply the SBME MEB by air. The proposed logistics glider is modeled as flown in a single tow sortie in Table 8. It is assumed that a logistics glider sortie has the least favorable tug orbit time of 32.5 minutes out of Table 14, "Towed Delivery Scenarios." Then each snatch pickup per sortie adds 15 minutes to the tug's orbit time, which was how long it took to switch towlines manually in WWII.[3] The other scenarios in Table 14 reduce orbit time, and automating the winch's switchover of towlines is expected to reduce scenario time even further.

Even without these performance enhancements, Table 9 shows other ways to use fewer tugs. Separating the tug from payload operations improves performance for even single snatch pickup sorties in SBME MEB resupply by air during an eight-hour shift. There is an even greater reduction in tug assets when conducting dual and triple snatch tow sorties.

The table also counts the tug assets necessary to retrograde during a later shift those reusable logistics gliders back to the Sea base. Note that this does not necessarily have to occur immediately after delivery but could be postponed depending upon any available stored logistics gliders and LZ security. Disposable logistics gliders will not require a retrograde option. The "VERTREP" column is the sling-carry method. The rest are by the snatch and stall technique. In the "Dual Snatch" column, the XG-22 and XG-23 are modeled with a field conversion to the XG-24 for a double return. Amphibious logistics gliders are released to the surface nearby the Sea base without requiring vertical return. This allows multiple amphibious logistics gliders per retrograde sortie. The drag on the tug for its capacity to physically tow multiple gliders is not assessed in this report.

Table 9. Single-Shift Tug Count

Air Delivery Option *Shift 1*	VERTREP	Single Snatch (tow craft)	Dual Snatch (tow craft)	Triple Snatch (tow craft)
MV-22	33	-	-	-
CH-53E	16	-	-	-
XG-21 Logistics Glider	-	22	16	12
XG-22 Logistics Glider	-	11	8	7
XG-23 Logistics Glider	-	7	5	4
Amphibious Log Glider	-	5	4	3
Retrograde Option *Shift 2*				
XG-21 Logistics Glider	(72) -	22	-	-
XG-22 Logistics Glider	(37) 19	11	6	-
XG-23 Logistics Glider	(21) 11	7	4	-
Amphibious Log Glider	-	5	3	3

The maximum assigned quantity of air assets to iHL aboard the Sea base has not been defined, but it is given that iHL should use fewer MV-22 or CH-53 as tugs than non-iHL all-air resupply approaches. In adding the delivery tugs and retrograde tugs, the sling carry sortie of a single logistics glider in the VERTREP retrograde method for SBME MEB quantities is shown (in parenthesis) not to be desirable. However, it may become practical when converting the XG-22 or XG-23 into an XG-24, which now halves the number of retrograde sorties as shown in the "VERTREP" column. Any non-disposable XG-21 scenario is completely undesirable without additional performance enhancement. The all-XG-21 and all-MV-22 were also undesirable from Table 8, but the remaining options are worth considering in SBME MEB sustainment.

Previously Table 8 modeled the XG-22 with a smaller typical per-vehicle payload than the MV-22, and the XG-23 being likewise smaller than the CH-53. Yet they require fewer tugs than the comparable VERTREP delivery method because cargo is processed independently of the tow vehicle. Only single tow XG-22 delivery and single XG-22 retrograde scenarios are less desirable than an all-CH-53 delivery; however, they are all more desirable than an all-MV-22 delivery. The remaining combinations of XG-22, XG-23, and amphibious logistics gliders use fewer tug assets than even an all-CH-53 delivery. The greater delivery range that logistics gliders offer is shown in Table 14, "Towed Delivery Scenarios." Scenarios of tug utilization, which take half as much time, are found there too.

There are combinations not recommended for iHL, and externally-supplied tugs have not been counted. Otherwise, the models presented here are a conservative representation that can be reduced with the other scenarios or more effective retrograde analysis. These tables show iHL as viable in Year 2015 SBME MEB daily ashore sustainment. It can be wholly met in an eight-hour shift using logistics glider variants at half payload capacity with fewer air assets on half of the 12 MPF (F) Squadron's helipad- or flight deck-equipped ships. The tow craft may be combinations of Sea base or non-sea-based joint

Navy, Air Force, Army, Coast Guard, and Coalition air tow assets, and this mix may fluctuate as needed during operations.

2.4.3 Small Unit Resupply

A key aspect of iHL is its scalable tactical resupply of MEB echelons down to direct resupply of small, remote, maneuvering units ashore. Centralized processing and convoys ashore are bypassed. A logistics glider does not put all resupply "eggs in one basket." While many logistics gliders may support centralized processing at some forward Airport of Debarkation (APOD), individual logistics gliders may also transit to distributed locations and be unloaded independently of the other logistics gliders' processing condition, location, or contents. The logistics glider may also be partially unloaded on a delivery route and then picked up for its next delivery elsewhere.

As a comparatively low value item, the logistics glider is well-suited for tactical delivery situations. It can approach the LZ silently, at high speed or high rate of descent, and land in restricted, unimproved areas without necessarily requiring ground combatant interaction or presence. High levels of autonomy minimize control and interaction by the recipient, or options can include sophisticated reprogramming or beacon-type landing guidance. The XG-21 and XG-31 are recommended for small unit resupply.

The legs of its tug are one of several factors in a towed glider's range to reach an LZ. Being independent of the logistics glider's performance, iHL is flexible in that the tug model itself may be changed for any number of reasons over the course of a resupply operation. As the distance to the LZ increases, inland specific tug models can be utilized as necessary for ceiling, fuel capacity, refueling options, or joint and coalition support.

2.4.4 Delivery Costs

There are many costs associated in delivery by general purpose vehicles and some of these are compared to iHL's tailored cargo infrastructure toward a high performance supply chain. The need for iHL could be debated given sufficient land bases, secure ports, shipping capacity, MAW assets, processing infrastructure, manning, and favorable sea conditions. But the purpose of the Sea base is to maneuver in an austere logistical environment where most of the friendly resources are constrained and are available only at a premium. While not a perfect solution (hence the interim aspect), the iHL concept has advantages over current air and surface delivery methods.

- Unit vehicle cost – without a pilot, engine, and fuel tank, there is a reduction in production complexity and the value from catastrophic loss.
- High value asset availability – there are fewer surface and ground cargo transfers required; and just the logistics glider may travel to the LZ. The tug is no longer tied up with loading and unloading.
- Deck parking space – the logistics glider is processed aboard the supply ship, stored in exactly its specified standardized footprint.
- Storage space – the logistics glider may remain ashore until queued for its next mission rather than take up any cargo space aboard the Sea base.
- Maintenance – without an engine and fuel, there is a reduction in preventative, diagnostic, and repair complexity.

- Manning - automation is encouraged, and without a pilot, engine, and fuel, there is a reduction in its supporting infrastructure.
- Fuel consumption – the tug snatches and releases multiple payloads in horizontal flight for a significant reduction compared to VERTREP fuel consumption.

A standardized container infrastructure is a key component in iHL design and performance, both inside and out. This encourages the needed automation around standardized container infrastructure such that traditional processing inefficiencies are reduced.

- Mixed cargos
- Strike up transfers
- Safety restrictions
- Repackaging processing
- Commercial handling equipments
- Reduced manning.

2.5 Tactical Lifecycle

The Sea base lines of operation are summarized from an iHL perspective, listing all possible functions. Then each stage of the logistics glider's operational life cycle is functionally decomposed by describing objective methods anticipated for transition to operations, as well as alternative approaches for design tradeoff consideration. The tactical sequence of these stages is in reverse order so that the end user comes first, and all system aspects funnel outward from delivery processing. Each stage describes the concepts and threat before the forward flow of key performance events guide the vision of operational vignettes. The employment environments are combined with a XG-22 logistics glider as the default vehicle. Additional tradeoff considerations are included to show options going beyond a minimal capability.

2.5.1 CAESR

The Sea base line of operation to Close, Assemble, Employ, Sustain, and Reconstitute (CAESR) provides an operational flow for many considerations when implementing iHL. The employment flow described next is based upon the previous sections assuming occupied flight and MEB-scale deployment, and pulls glider tactics from WWII.

A Marine Expeditionary Brigade has closed and assembled in a Sea base off the enemy coast. The sea-based logistics gliders are densely stowed, awaiting unpacking. Supply routes will have air superiority. Upon employment of forces ashore, logistics glider preassembly and preload begins in preparation for the sustainment phase. Each cargo vehicle could be loaded with any combination of the following:

- Dry cargo in various containers such as JMIC, ammunition, and repair parts
- Wet cargo:
  - o Petroleum, oil, lubricants (POL) and packaged water for unload
  - o Tanker configuration for water distribution or refueling stations

- HMMWV, up armored
- Howitzer, LW155

Surge delivery then occurs by any combination of the following three iHL resupply methods:
- Glider snatches from supply ship helipads with minimal weight and low value payloads if only minimal options for launch failure exist
- Either glider snatches or towed by short-takeoff tugs from ship runways with intermediate weight or higher value payloads as launch failure options permit
- Conventional towed glider takeoffs from runways at advanced supply base(s) with maximum payloads and non-time-critical payloads

Multiple logistics gliders may be towed by one tug, and each terminates for one of the following:
- Free flight to the same LZ
- Free flight to separate LZ
- Disposal of its payload during tow and retrograde when within VERTREP weight specification

After release, a logistics glider's towline is handled in one of the following ways:
- Dropped
- Dangled to the next snatch
- Reeled into the tug

The tug may perform the following after releasing a logistics glider:
- Travel to the next waypoint
- Release a remaining towed logistics glider
- Return to the Sea base
- Standby for its recovery
- Snatch another waiting logistics glider
- Perform other non-tow operations, depending on mission configuration

Logistics gliders will have imprecise descent paths that vary due to weather and vehicle performance. Airspace with one or more restricted maneuver capability vehicles will need collision-avoidance procedures built into the vehicle. Potentially dozens of air vehicles may arrive at semi-random intervals to the following:
- Waypoint
- Approach pattern
- Landing zone
- Ground taxi area
- Parking space

Ground operations upon landing may include one or more of the following:
- Wait
- Tie down

- Be pulled a short distance by ground vehicles
- Have the bottom towed as a flatbed a short distance by ground vehicles
- Unload or partially unload
- Disposal
- Load with:
  - Retrograde equipments
  - Medical evacuation
  - Troops for reconstitution to the Sea base
  - Troops for movement within the JOA

Upon completing ground operations, the logistics glider is either horizontally snatched or vertically sling-carried away toward the following:

- Retrograde to a Sea base supply ship
- Retrograde to a Sea base runway
- Retrograde to a shore depot
- Retrograde to an advanced base runway
- Continue to another LZ for performing:
  - Partial unloads
  - Troop movement

Once returned to base, the following retrograde operations are performed:

- Reconstitution
- Repairs
- Reconfiguration
- Replacement
- Preflight certification and reload

2.5.2 Ground Operations

The customer comes first. The supply chain funnels inward to the user effectively consuming only needed materiel in a timely manner with minimal expenditures, processing, waste, and overhead. Considerable design effort must be given to the interaction by ground personnel—who are likely unfamiliar with handling air vehicles and their safety procedures—to avoid damage or injury. This includes robust logistics glider materials and ergonomic design, and multiple levels of training.

A hypothetical vignette of a small unit resupply operation starts with an infantry squad securing the designated LZ. It is an open field and a XG-22 logistics glider arrives autonomously. Since it is known to be heavily loaded and the squad lacks equipment to move it, planners intentionally selected an LZ with enough takeoff clearance that the logistics glider does not have to be repositioned.

The glider is accessed by two teams for the rearmost two JMIC having supplies. One team empties the container providing all resupply items: water, MRE, ammo, batteries, etc. Fire squads rotate by to load up. Any returning items are secured inside a retrograde container, such as used batteries, trash, gathered intelligence, etc. A depleted JMIC is folded up and stowed inside the logistics glider.

Meanwhile the other container holds the towline and all necessary ground station setup gear. Its setup checklist is posted conspicuously and followed by the other team.
1. Pace off an "L" from the nose of the glider to where the station will go.
2. Pound in a stake and assemble one station pole, inserting it on the stake.
3. Assemble the other station pole and expand the integral towline loop until it is tight.
4. Pound in the other stake and insert the pole on it.
5. Unroll the towline back to the nose of the glider and wait until the glider is ready for flight.
6. Clear all personnel from the flight path and only then attach the towline to the glider.
7. Signal the tug for pickup.

The squad holds the LZ secure until the tug snatches the logistics glider away.

Threat considerations against ground operations include weather, terrain, hostile actions, handling, and mechanical failures. Weather threats experienced will be similar to those of other air vehicles in the JOA. As they are unoccupied, logistics gliders are more likely to be grounded overnight or left exposed to severe weather requiring a secure tie-down. Gliders are particularly sensitive to high winds, and depending on construction and skin material, severe weather conditions such as hail damage might also be a concern. Many undamaged WWII gliders were lost to post-invasion storms. Some suggested procedures include:
- Stow aboard each logistics glider sufficient tie down gear to stake lines from the standard four top corner TEU handling points into the ground.
- Pop off the outermost wing tip sections to reduce wing operation in gusts and to protect those control surfaces. In this situation, handling and storage concerns by inexperienced personnel will need to be addressed.
- Disassemble the tail section for the same reasons and cautions.

The logistics glider, of course, could always be disassembled, but there may not be an empty TEU and appropriate packing material around to store the sensitive flight surfaces. A single cargo bay cannot hold all wing and tail sections. The nose cone should be impervious to hail damage if it is designed to withstand towline breaks and collisions with landing obstructions.

Terrain is a big factor in considering damage to the logistics glider. Given the types of onsite equipment available, consideration must be given to completing its mission and recovering the logistics glider after a hard landing or coming to a halt in undesirable positions or locations. Some training or design issues include access to the payload under a collapsed airframe, jammed doors, and other hard landing side effects.

Consequences of hostile actions would be similar to those for other air vehicles on the ground. Due to being unoccupied and potentially abandoned for periods of time, logistics gliders may suffer access by non-friendly agents. Anti-tamper functions should be available with simple procedures on implementation and disabling. The sides will likely provide little protection to any hostile actions. The nose cone and then the bottom will have the thickest skins.

2.5.2.1 Offload.

Access to the newly landed logistics glider must require only tools and training commonly available to ground forces and western military culture.
- Anti-tamper procedures must be commonly identifiable to untrained personnel, such as keypad entry to disable cockpit-housed security measures.
- Ramp/doors/jacks
- Spine rail(s)
- Interface to ground vehicle winches in the front and rear
- Vehicle and personnel obstructions (wing height, entangling lines, etc.)
- Mating up to a MTVR flatbed for cargo transfer
- Availability and applications requiring cranes and manning

The offload of heavy bulk cargo using the body-on-frame design suggests the following generalized sequence of events:
- Disconnect glider and cargo flatbed
- Jack one of them up for separation
- Push or tow the glider and cargo flatbed apart
- The cargo flatbed is raised to the height needed for cargo processing, such as level with the bed of an MTVR.

Once reassembled, towing on the ground for short distances requires at least the following hardware:
- Structurally sound locations for hand placement (pushing, pulling)
- Ground tow hitch or tow hook connection in front
- Hook for rear connection to a ground vehicle winch
- HMMWV-compatible rear tires, 37 in. in diameter x 12.5 in. wide (Note that R16.5 LT load range "D" radial wheels are limited to 3,850 lbs capacity each.)

2.5.2.2 Ground Intercept.

The reusable aerostat for at-sea intercept is likely too expensive and unnecessary for single-use ground snatch. A disposable balloon version may provide an intercept station for ground-based logistics glider snatch. This may reduce the risk to the tug during snatch pickup in a hot zone.

Much less complex is duplicating the WWII ground intercept station as described in the WWII training film.[7] The appropriate tools will need to be supplied to put the stakes in the ground when erecting the station poles. Ground setup and coordination with the tug for glider snatch pickup may require specialized training and procedures.

The previous hypothetical vignette is expanded to include the tug's perspective for ground intercept. The tug had released the glider earlier and monitors the surveillance video provided during the logistics glider's approach and landing. Any additional surveillance is checked with other logistics gliders in flight in the vicinity. Ground radio contact ascertains that the logistics glider is ready for pickup and the LZ remains secure. A meteorological broadcast is received from the logistics glider, and the tug's approach vectors are verified. The tug flies a single pass over the ground station to visually verify the setup, wind conditions, obstructions and clearances. It then circles around for snatch pickup. The intercept at the ground station occurs at high speed and then the tug climbs out with the logistics glider in tow.

The ground station poles can be integral to a towline's contact loop. The tube sections are unfurled and slide together into one pole. The contact loop is formed by stretching apart the two poles. They are mounted to stakes hammered into the ground. If the stakes are designed and installed properly, the logistics glider may take off directly through where the ground station was installed. This may simplify ground setup procedures by eliminating pacing off the "L," but the safety clearances between the tug, its boom, and the logistics glider will require practical evaluation.

2.5.2.3 Retrograde Qualification.

The logistics glider will need preflight inspection, launch positioning, and towline rigging prior to departure out of the LZ. As per WWII procedure, the tug will require the weight of the logistics glider prior to snatch pickup.

2.5.3 Descent

The tow and descent stages have similar threats (weather, hostile action, collision, and equipment failure modes). The impact upon navigation during free flight from varying wind vectors at different altitudes will need consideration.

All threats are amplified closer to the ground, with landing modeled as an intentional collision. Terrain obstructions and surface topology, density, and wetness are big factors in damage to the logistics glider with its wingspan, stopping distance, and unoccupied landing mode. An unimproved LZ has many hazards to a rolling airframe, including other vehicles. It is suggested that the outermost wing tips pop off easily upon any impact, while the nose cone and bottom be very strong to survive impact with fences, scrub, and modest walls.

2.5.3.1 Release.

All gliders decelerate immediately upon release from tow. The release maneuver for a recreational glider involves a climbing turn to the right and the tow plane diving to the left for maximum safety separation. The preferred recreational tow position is the high tow so any impact damage from the released tow hook occurs to the bottom of the glider.

Similar procedures should be considered in iHL. Avoiding collisions with other towed vehicles and all operational environment considerations need to be taken into account.

Obviously the tug should not maneuver violently with remaining gliders in tow. The low, towed glider should dive rather than climb upon release to avoid contact with the towline.

2.5.3.2 Free Flight.

Once clear of the tug, towlines, and other towed vehicles, the glider is in free flight mode. There are likely to be primarily two free flight profiles: either at an optimal glide slope to travel a far distance to a waypoint, or at the redline speed for greatest survivability and accuracy. Defensive maneuvering can be preprogrammed against expected threats at specific waypoints, but there may be tactical or cost limits to sufficient sensing capabilities or communications to react to all threats.

A glider's lift over drag ratio is its rated glide slope traveled in calm air and is independent of air density and weight. Its velocity changes with its weight. The lift prefix number is the distance it will glide to an LZ given the drag postfix number for the starting altitude. Glide ratio is an internally designed specification; the actual glide path will vary outside the ideal speed range and with changing meteorological conditions such as the following:
- Thermals
- Ridge lift
- Mountain waves
- Leeward rotors
- Downdrafts
- Storms
- Icing
- Jet stream, headwinds, and tailwinds

The flight mode used in WWII was to deliver the invasion glider to the LZ and then release it for as rapid and vertical a descent as possible. The CG-4A had a poor glide ratio before considering the significantly overloaded conditions reported in several operations. The CG-10A was rated at 14:1 so perhaps a high altitude, standoff release for a volplane delivery was anticipated. The primary modern consideration is likewise to survive delivery with minimal arrival errors. Optimal glide slope is a secondary consideration.

A volplane flight profile nearest the logistics glider's optimal glide slope should be considered to minimize tug utilization, fuel, and JOA exposure. There are risks with headwinds, downdrafts, weather events, and exposure to hostile action. The typical flight speeds for the optimal lift over drag ratio is found nearer the slow end of the flight envelope rather than redline speed. Redline speed is the safety margin in aircraft integrity. However, there are design approaches to increase glide ratio slightly toward the faster end of the speed spectrum. Additionally, weather conditions such as tail winds, ridge lift, mountain waves, and thermal activity can also significantly improve glide slope over the rated glide ratio. Table 14, "Towed Delivery Scenarios," shows several delivery scenarios given calm air.

2.5.3.3 Targeting.

Autopilot technology in targeting delivery to a waypoint in space is technologically mature. It is mainly a question of size, power consumption, weight, and cost of the computer and sensing technology. Complex autonomous flight control is maturing. However, the entire field of autonomous military platforms making integral decisions without direct human oversight can benefit from an open architecture standard in command and control. iHL provides a basis to begin such an effort.

The low flight speeds of the logistics glider help reduce the decision-making complexity. It is suggested that, as approach control errors become a factor near waypoints, the logistics glider fly a maximum descent rate path. Radical maneuvers beyond flaps and spoilers, such as slips and flat spins can be very effective for descent but are inherently risky and not yet mature in autonomous air vehicle control.

External navigational guidance can be passive sensing such as GPS, or homing beacons provided by forward controllers. Active guidance technologies include sonar, lidar, or radar. Active guidance communications may need to include a wave-off feature and integral logistics glider decision-making when active guidance is not received as expected. This implies alternate LZ or suicide options need to be preprogrammed in the event no external directives are received by the logistics glider.

2.5.3.4 Landing.

Glider approach into an LZ will be different than, but built upon, vertical insertion tactics learned since WWII invasion glider tactics. Scenarios need to be categorized into logistics glider capabilities. There will be options for short field landings versus having a long unobstructed roll. The parking and processing plan will also impact the landing techniques and procedures chosen.

Not recommended is a WWII CG-4A landing technique when coming in short: the pilot would slam the glider down early and hop over obstructions such as hedgerows and telephone lines. Although claimed to take advantage of the suspension, likely the steel tube and canvas construction of the CG-4A contributed to the flex and spring action. This would not be guaranteed with modern materials or constructions. A safer approach would be to accelerate to the deck and utilize ground effect as the speed bled off.

The logistics glider deceleration technologies start with the reuse of onboard launch capabilities such as wheel brakes and flaps as air brakes. Many slowing and stopping lessons were learned in WWII glider test and experimentation, and many more have been developed since. Unsuccessful experimentation during WWII included drogue chutes, plows, and even retrorockets. The combination of nose skid, wheel and air braking proved the most robust under some arctic conditions.

Since WWII, many of the unsuccessful techniques have matured. High impact Indy barrier technology has recently dissipated velocities in the flight range of the logistics glider. More investigation is recommended in this area.

2.5.3.5 Parking.

Given modern terrain survey capabilities and advanced planning technologies, the organization of parking and storing many logistics gliders at one LZ can be detailed well in advance. LZ software can plan the efficient storage, unloading, reloading, and even repair of logistics glider assets on the ground. By keeping inventory at secure ground locations, this alleviates the footprint issue on the Sea base by only recalling logistics gliders when the ships are prepared to receive and process them.

Given an advanced parking plan, the inertial taxiing of the logistics glider on the ground upon landing can be aimed at a final waypoint programmed into its mission. Alternatively, internal navigation technology or ground personnel with secondary parking beacons can redirect the taxiing logistics glider off the landing zone to a parking area. Wheel and air braking technologies offer control of the logistics glider when it is no longer flying but still maintaining its inertia.

Ground vehicles and crews may be needed for logistics glider parking operations, including final planning, tie downs, safety zones during unloading, clearing the LZ, preflight inspections, and positioning for launch.

2.5.4 <u>Snatch and Tow</u>

In the littoral, the iHL system includes the tug and its iHL equipments, and the details of snatch launch of the logistics glider off the deck.

It was WWII procedure to receive the weight of the logistics glider prior to snatch pickup for approach speed and winch settings. Often the tug would perform multiple preparatory passes over the intercept station to determine crosswinds, achieve better radio contact, or visually determine the progress in launch preparation and read the signal markings up close.

2.5.4.1 Tug.

There are many tugs available to perform runway-based tow, ground snatch, and helipad snatch. Some tugs may have performance sufficient to support some but not all iHL functions such as multiple snatch pickups, delivery to a glide altitude, delivery to an LZ, retrograde, and in-JOA delivery and pickup routes. Accounting for the forces of helipad snatch will provide performance criteria to help reduce this pool of tug candidates.

The installed equipment necessary to perform glider snatch may also reduce the tug pool. The primary modification is in adding the winch and boom, but there are also structural mounting points, power, significant heat dissipation, and the dynamics of towline position during snatch and tow. Tail hook-equipped carrier aircraft may work. Joint and coalition forces may also provide tugs for a straightforward interface to the Sea base, but this report only models USN and USMC air vehicles.

The historical experience supports propeller-based tugs, which the Navy has in the P-3, MV-22, and C-130. All generally meet the logistics glider's flight performance range. The P-3 may be out of service by 2015. The MV-22 is the favored candidate, with its tilted proprotors capable of providing an off-axis thrust during snatch pickup, which helps keep the towrope away from its elevator. Even tilted, its long propellers may require a longer pickup boom and related safety margins during intercept. The C-130 needs a land-based runway for takeoff and landing, but with sufficient planning it can operate in the JOA given its long range and mid-air refueling capability to increase range.

Any runway-based towing must take into account a quick transition to flight for the logistics glider compared to the tug. As a safety concern for recreational glider towing, this can easily pull the tug's tail up. This is exacerbated by the logistics glider's great mass and low rotation speed. Especially with multiple gliders in tow, a tug will have considerable drag behind it and hence spend more time on the runway reaching its rotation speed.

Potentially, the rotorcraft inventory may support iHL at different performance points. A rotorcraft does not have the same stall speed minimum requirement that fixed-wing aircraft must avoid, so more momentum can be bled off into the intercept than with an airplane. The CH-53E cruises between 120 to 150 knots, which is in the acceptable range for logistics glider snatch and tow. From experiences with rotorcraft intercepting parachutes in mid-air, the rotor downdraft is an issue to monitor; however, snatch speeds should be high enough for it not to be an issue.

Jet aircraft are not excluded from iHL, but there needs to be some supporting analysis for the winch placement, the candidate aircraft's minimal speeds and performance ranges, exhaust heat, and overall fuel savings. There are the P-8 and C-17 to potentially evaluate.

2.5.4.2 Winch.

The winch places requirements upon the tug such as volume, power, heat dissipation, structural mounting, and towline movement zones out of the tug. The winch unwinds cable down the boom in preparation for snatch contact. Upon contact, the winch pays out the cable on its drum with careful increases in line tension by braking. Once the logistics glider becomes airborne, the drum winds in the excess towline.

It is unsafe to permanently attach the end of the towline to the winch drum. In emergency situations during intercept or tow, the tug cannot be restrained by an inadvertent towline collision. It is best to let the line pay out of the winch as the tug escapes from a dangerous situation. Not permanently attaching the towline implies the logistics glider can be lost if the towline pay-out length is exceeded.

For repeated snatches during one sortie, the active towline must be automatically switched off the winch drum. The winch must be designed to process several different towline scenarios.

2.5.4.3 Boom.

The boom is the exterior addition to the tug, extending for contact only during snatch intercept and kept retracted otherwise. It must be long enough to keep a significant safety margin between the intercept station and tug's propellers, landing gear, etc. It must be sturdy enough to survive impact with the tensioned loop yet not cause the loop to break at contact velocities that are likely to be higher than WWII experience.

Once extended, the boom runs the tow hook from the winch, out of the tug, and down to the tip of the boom. Upon contact, the boom slides the towline loop down its length and into the hook. The hook engages the towline and separates from the boom. The boom is then either retracted for efficient flight or reloaded for the next snatch.

2.5.4.4 Towline.

The towline is a complex combination of steel and feeder cables, weak links, twisted nylon line, swivels, thimbles, a contact loop, and tow ring. Towlines will be exposed to all the environmental elements at sea, in the air, and on the ground. Nylon especially has lowered breaking strength when wet. Different towlines may run up to an aerostat, ground contact station, or act as a tether for a disposable balloon. There is evidence that WWII glider snatch towlines transitioned to a thicker nylon section for more reliable or heavier snatch operation. Potentially different towlines or towline components may be involved for different glider weight ranges or towing scenarios.

Towlines were reused during WWII development by dropping them back over the airfield. In combat they were disposed of after glider release. A disposable iHL towline configuration might use the following assembly in series from the glider to the tug:
1. Glider tow ring
2. Elastic nylon
3. Steel feeder loop
4. Intercept hook
5. Steel payout cable

It is possible to reuse the towline in iHL. This implies a different towline assembly and affects the iHL designs for the winch, tug cargo volume, and snatch tow scenario details. The nylon section reels in and pays out of the tug. The tug gains a steel feeder line after each snatch pickup, which will have to be disconnected after each release, accounted aboard on another drum, and returned at the end of the day. A reusable iHL towline configuration might use the following assembly in series from the glider to the tug:
1. Glider tow ring
2. Steel feeder loop
3. Intercept hook
4. Elastic nylon
5. Steel payout cable

2.5.4.5 Launch.

Table 10 lists the Newtonian force and, given constant acceleration, the kinematic equations for the roll across the helipad prior to logistics glider liftoff at rotation speed into towed flight.

Table 10. Force and Kinematic Equations

$F = ma$	Newton's 2nd Law	(Equation 1)
$x = x_0 + \frac{1}{2}(v_0 + v_x)\,t$	Without a	(Equation 2)
$x = x_0 + v_0 t + \frac{1}{2}a\,t^2$	Without v	(Equation 3)
$v_x = v_0 + a_x t$	Without x	(Equation 4)
$v^2 = v_0^2 + 2\,a\,(x - x_0)$	Without t	(Equation 5)

The assumption that the snatched glider accelerates at a constant rate requires further exploration. The measure for helipad snatch is the acceleration a to flight rotation speed v_x in the distance x across a helipad.

Table 11 uses Equation 5 to calculate the acceleration in gravity force (G) needed to travel a distance in feet to achieve CG-10A rotation speeds between 62 mph (53.9 knots), 50 mph (43.4 knots), and down to the speed necessary for 0.70 G acceleration (30 mph). This distance traveled is not greater than the helipad dimension minus the wheel footprint of the logistics glider.

Table 11. Gee Forces to Rotation Speed Given Rollout Distance

Rotation	53.9 knots	50.4 knots	46.9 knots	43.4 knots	40.0 knots	36.5 knots
77 feet	1.6 G	1.4 G	1.2 G	1.1 G	0.9 G	0.8 G
79 feet	"	"	"	1.0 G	"	0.7 G
81 feet	"	"	"	"	"	"
83 feet	1.5 G	1.3 G	"	"	0.8 G	"
85 feet	"	"	1.1 G	"	"	"
87 feet	"	"	"	0.9 G	"	"
89 feet	1.4 G	1.2 G	"	"	"	"
91 feet	"	"	"	"	"	0.6 G

The 77-ft takeoff distance is an approximation of the distance available to a CG-10A placed on a 32-meter-wide helipad. From Table 19, "Helipad Snatch Model," a logistics glider would have 89 feet available to launch from the helipad.

Any retractable wheels must not fold upon liftoff while there is the possibility of contact with the helipad or ship. Then they should fold before there is any possibility of contact with the water surface.

2.5.4.6 Airfoil.

The wing and body lifting force and GVW equations are provided in Table 12. Equation 6 presents C_L as the lift coefficient, q is the dynamic pressure, and S is the wing lifting surface area of the glider airfoil (the wing and a lifting body of the same chord as the wing root). This equation is not calculated in this report, as design values for C_L and q have not yet been determined. Rather Equation 7 uses a linear proportion model where W_{load} is the CG-10A wing-lift rating in pounds per square foot of lift area. Gross vehicle weight is assumed to be equivalent to or less than the lifting force L.

Table 12. Lifting Equations

$L = C_L q S$	(Equation 6)
$w_{load} = GVW_{CG-10A} / S_{CG-10A}$	(Equation 7)
$GVW = w_{load} S$	(Equation 8)

The lifting force is proportional to a glider's wing surface area. Using the CG-10A design as a baseline for a first-look model, its 1180 sq ft wing surface was specified at 21.2 lbs per sq ft. Yet the CG-10A was demonstrated to lift 32,000 lbs gross weight for a lifting force to range up to 27 lbs per sq ft. From its empty weight of 12,000 lbs, this payload capacity range is presented in Table 13.

For the logistics glider variants listed in this table, it is assumed there is no lifting surface area increase via variable geometry or unfolding its stored wing components. Based on the maximum square footage that can be packaged into an ISO standard TEU container, the logistics glider wing lifting surface area is derived from its wing components in Tables 6 and 7, "Logistics Glider Assemblies." The linear proportion model uses Equation 8 to provide the estimations in the "GVW range" column. This might not be precise or even applicable to non-wooden gliders. "Payload" is the net vehicle capacity range between 49% and 62% of GVW.

Table 13. Airfoil Lift Capacity

Glider	Lift area (sq ft)	GVW range (lbs)	Payload (lbs)
CG-10A	1180	25,000 - 32,000	11,000 - 20,000
XG-21 Logistics Glider	418	8,900 - 11,300	4,300 - 7,000
XG-22 Logistics Glider	834	17,700 - 22,600	8,600 - 14,000
XG-23 Logistics Glider	1548	32,800 - 42,000	16,000 - 26,000
XG-24 Logistics Glider	1548	32,800 - 42,000	16,000 - 26,000
XG-31 Strap-on Glider	418	8,300 - 11,300	4,300 - 7,000
XG-32 Strap-on Glider	834	16,600 - 22,600	8,600 - 14,000
XG-33 Strap-on Glider	1548	32,800 - 42,000	16,000 - 26,000
Amphibious Log Glider	1890 - 3250	51,000 - 65,000	32,000

It is assumed (albeit poorly) that the bi-wing configuration scales its lift efficiency linearly to a mono wing. This is not typically the case and requires an advanced design effort to develop a better model.

The amphibious logistics glider is a non-standard ISO shape and as such is not constrained in volume by its packaged shape. So its lift area and GVW are estimated backwards from the 60% payload capacity model filled with a large vehicle. The heaviest MTVR option is chosen because it is slightly heavier than the LAV.

2.5.4.7 Flight Control.

Ideally autonomous flight control is handled by the logistics glider's decision-making algorithms. Most glider flight functions are not complex, and autonomous decision-making can be designed with passive monitoring. Information preprogrammed before flight should at least include default contingency operation. Much other information may need to be preprogrammed depending on navigation and control technologies selected. The issue of in-flight reprogramming will need to be addressed.

2.5.4.8 Multiple Snatch Pickups.

In the few WWII-era dual pickups demonstrated, each glider assumed a high tow position, with the first yawing away during the snatch of the second. Both assumed the high tow position for clearing any ground obstacles, which is not an issue for balloon intercepted snatch pickups. There has been no attempt at a triple snatch.

The WWII dual pickup demonstrations transferred an active towline off the winch in preparation for the next snatch pickup. It took about 15 minutes to complete. This is too complex and dangerous for manual application. It will need to be automated as part of the iHL winch system aboard the tug.

iHL proposes two or three logistics gliders snatched in succession by one tug per sortie: the first and any middle glider will assume a high tow position. The high tow position is preferred to give the greatest safety margins to clear shipboard obstacles. Only the last logistics glider snatched in succession will take the low tow position so as to maximize towed glider separation.

2.5.4.9 Towed Flight.

After snatch pickup, the logistics glider is in towed flight mode. In WWII an automatic pilot was engineered and approved during experimentation at CCAAF.[10] This was because glider pilots became disoriented in clouds and failed to maintain correct tow position and safety margins. It was not operationally accepted before the end of WWII. There were technical issues with the elasticity of the towline and with tug maneuvering during combat. Modern autopilot technology is expected to have less risk in this regard.

The release of the logistics glider affects the utilization of tug as a launch asset. Table 14 uses conservative estimates for simple travel models toward the LZ after snatch pickup. Five scenarios show the tug's round trip time and glider delivery range in nautical miles (NM) to an LZ within the JOA. Scenario 1 is a simple time and distance estimation of a level tow to the LZ. The logistics glider is towed at redline (maximum) speed and released at 500 ft above the LZ, while the tug returns at cruise speed. The remaining scenarios have varying flight profiles by towing at a steady climb rate with logistics glider release upon reaching the prescribed altitude in feet above sea level (ASL). The glider volplanes to the LZ at a speed for best glide performance, typically significantly less than redline. Depending on its glide ratio, the glider travels a total distance in nautical miles away from the Sea base to the LZ also at 0 ASL.

Table 14. Towed Delivery Scenarios

Scenario Example	Tow (knots)	Cruise (knots)	Climb (ft/min)	Release Altitude	Round Trip (min)	Glide Ratio	LZ (NM)
1	150	240	0	500	32.50	-	50
2	150	240	1,000	10,000	16.25	15:1	50
3	150	240	2,000	10,000	8.13	15:1	37
4	150	240	2,000	20,000	16.25	15:1	75
5	150	240	2,000	20,000	16.25	22.5:1	100
6	150	240	2,000	24,700	20.07	30:1	154

Scenario 3 increases the tug's climb rate of Scenario 2, which reduces the tug's round trip time. For extending delivery range, Scenario 4 increases the release altitude and Scenario 5 then increases the glide ratio. While the tug's engine performance and hence climb rate is affected by thinner air (which is not modeled in the table, only the overall average climb rate), a glider's lift over drag (glide) ratio in calm air always remains the same regardless of its weight or air density altitude. This is an advantage when trading options on delivery throughput performance or asset availability.

Scenario 6 climbs to the maximum ceiling for the MV-22 and releases a logistics glider on a favorable performance glide slope. While top racing sailplanes double this glide ratio and even dual place sailplanes rate 38:1, a logistics glider is an order of magnitude heavier and may not intrinsically rate this level of performance in still air. However, it is typical to take advantage of favorable weather conditions with good forecasting and planning to increase the gliding range. Again an average performance indicator can be misleading and inappropriate to every specific situation.

The volplane scenarios 2 through 6 may be easily modified for greater delivery ranges. The tug tows additionally fly farther toward the LZ without changing altitude. For example, adding a delivery distance increase of 50 NM, the round trip time from Scenario 1 can be added to another scenario. For multiples of 50 NM, add multiples of Scenario 1.

2.5.5 Basing

Appendix A lists the complete naval equipments and hull population anticipated in year 2015 to consider for helipad snatch. This includes the following.

- Launching ship class with available helipad dimensions and potential for converting to iHL. The length of the helipad is the aft-to-stern distance between obstructions over 5 in. in height. The width is usually the beam of the ship or available space to park and launch the logistics glider. The height is above the specified full waterline.
- Tug class and relevant specifications
- Standardized payload and relevant specifications

2.5.5.1 Advanced Base Delivery and Assembly.

The Advanced Base is an upstream link in the Sea base supply chain. Advanced Bases have friendly ports and runways for the embarkation of expeditionary cargo. With midair refueling there are two, not necessarily exclusive, Advanced Base methods iHL may explore.

In the first method, Joint and Coalition tow craft bypass the Sea base. The bulk of preplanned, forecasted supply materiel is towed from the runway of Advanced Base(s) for standoff release into the JOA. In either surge or chainsaw waves, only logistics gliders penetrate the high-risk zones. Logistics gliders are either preprogrammed by launch, or redirected upon arrival for navigation to the LZ.

The second consideration uses Advanced Base tow assets for Sea base helipad snatch: arriving above the Sea base, picking up, climbing to a standoff release, and then returning while the logistics gliders perform their precision delivery mission.

The Advanced Base logistics gliders may join and intermingle with Sea base logistics glider inventory during sustainment and reconstitution phases. Advanced Base logistics gliders are recovered by two alternatives. Sea base rotorcraft recover logistics gliders for littoral processing as covered in this report. Otherwise the Advanced Base tugs may enter the JOA to ground snatch and return inventory to the Advanced Base. Either way this reduces the sea-based transport and storage requirements.

Under specific conditions during the Sea base sustainment phase, using Advanced Base assets reduces the volume of sea-based cargo needlessly participating in the closure, assembly, and employment phases. It provides forecasted or preplanned resupply but not rapidly changing delivery locations or custom requests, which is a Sea base strength.

2.5.5.2 Sea Base Delivery and Assembly.

The logistics glider is brought aboard either as cargo during portside loading or VERTREP during operations. It is stored as a multiple of TEU as cargo. Unpacked for use, its body will proceed to a cargo load area before final assembly with its wings and tail.

Performance modeling reveals the Sea base cargo strike up and repackaging processes as the longest and most inefficient links in the Sea base supply chain.[1] iHL provides the opportunity to automate and standardize this processing for significant performance improvement and manning reductions. The ideal iHL implementation has a shipboard automated warehousing system selectively providing factory-prepared JMIC directly to logistics glider loading area. Using sea state-insensitive technology, this payload is loaded aboard the logistics glider. The sequence of loading is based upon the unpacking sequence after delivery, center of gravity limitations, restrictions from any hazardous items, etc.

The contents of the JMIC are typically a factory prepared, standardized mix of resupply materiel designed for immediate use by the designated expeditionary unit. Rapidly requested items are a smaller percentage and require some manual involvement depending on the level of automation in the decision support software described next.

2.5.5.3 Decision Support Software.

The planning and execution of many aspects of iHL operations encourages a level of decision-making and forecasting software tools not yet envisioned for expeditionary logistics chain management. Improvements in logistics control and inventory bandwidth, database interactions, and automated decision support will allow very effective asset allocation, efficient delivery planning and replanning, reduce loss and waste, and require minimal manning for oversight. Interacting with automated warehousing and assembly, this can significantly reduce Sea base manning.

2.5.5.4 Retrograde Delivery.

The iHL concept is conservative to the point of inefficiency in the return of the logistics glider to the Sea base. There are many alternative methods to arrive at the Sea base such as landing upon a speeding 40-knot craft, arrestor wires, drogue chutes, L-deck runways, and retro rockets.

The proposed method is to vertically deliver the logistics glider to the Sea base as a sling-load under a MV-22 or CH-53. There are two techniques for the return flight to reach VERTREP at the Sea base. The simplest and least efficient is to VERTREP the logistics glider as a non-flying, sling-carried payload the entire distance from the LZ to the Sea base. This is essentially a downed aircraft salvage operation, only designed integrally as part of the iHL system. This is also a damaged logistics glider retrograde mode as the majority of incidents will be to its wings or bottom, preventing the proposed method described next.

A novel but much more efficient retrograde approach is to snatch and later stall the logistics glider for VERTREP delivery. This may have tactical appeal in that the tug's speed stays high within the JOA, reducing its vulnerability. Retrograde operation is performed as follows.

1. The logistics glider is either rigged at pickup with a special three-point transition harness, or this harness system is integrally designed into the logistics glider. This harness connects to the nose towline and to the two rear ISO lift points.
2. Snatch pickup is performed by a tug capable of VERTREP of the logistics glider's dead weight.
3. Towed flight occurs until arrival in the vicinity of the Sea base.
4. Over open water with appropriate safety margins, the tug slows to just above the logistics glider's stall speed and intentionally stalls the logistics glider gently into a sling carry position underneath. The onboard winch can participate in this stall and transition.
5. VERTREP occurs onto the Sea base deck.

2.5.5.5 Delivery Inspection.

The logistics glider will need delivery inspection upon return to the Sea base for any repairs, continued flight suitability, etc.

2.5.5.6 Packaging Handling Storage Transport (PHST).

The iHL objective is to package the logistics glider into multiples of ISO standard TEU footprint units, complete with standard contact points for storage and transport. Supply ships process this configuration in a Sea base resupply system. With ISO standard TEU handling points, mounting brackets, and additional supports, they can be brought aboard and stacked on top of each other.

The following assembly steps can be performed below decks or separately above decks depending upon space available and the area to store partially completed assemblies. The first stage is to unpack and assemble the fuselage body, then the six wing segments for final assembly.

1. Attach nose cone to body.
2. Attach any skid to nose cone.
3. Attach tail boom.
4. Attach rear doors.
5. Attach left and right leading wing edge (includes any spoilers) to body.
6. Attach left and right trailing wing edge (includes flaps) to body and leading edge.
7. Snap on left and right wing tips (includes ailerons) to assembled wings.
8. Lift up rear using built-in jacks, tilting forward.
9. Push in flatbed bottom, with or without payload.
10. Perform final assembly and flight checks.

2.5.5.7 Payload Preparation.

The Joint Modular Intermodal Distribution System's concept for the Joint Modular Intermodal Platform (JMIP) needs to be advanced into a structural, detachable, aerodynamic logistics glider bottom. The JMIC is the smallest preferred unit to load aboard the logistics glider.

In most cases, standardized cargo units need to be prepackaged, tracked, and identified to facilitate automated retrieval and assembly. Cargo is loaded onto the detached bottom by automated equipments below decks. Then JMIP-compatible forklifts or towing vehicles transport the bottom to the glider assembly area. The logistics glider fuselage is attached to the preloaded bottom and the flight-worthy vehicle is towed to launch position.

2.5.5.8 Aerostat Processing.

The concept of a tow craft snatching a loaded supply glider off a supply ship requires the lofting of the towline clear of ship rigging, and that it is stable in regard to sea motion. There would be preparation, lifting, and reloading periods. The preferred concept would loft a reusable aerostat 500 to 1000 ft ASL from a supply ship. This zone is typically stable for air buoyancy technology although less so in the littoral than other regions.

In no-wind conditions, the aerostat's payload must be supported completely by air buoyancy, not by any aerodynamic lift. The payload consists of sensors (wind speed, direction, accelerometers, collision detection, and communication to the deck and tow craft), the aerostat's anchor tether, and part of the glider's towline. The design of the aerostat must primarily minimize z-axis motion, especially in the positive direction as the tug approaches and intercepts the towline.

Of secondary preference is a larger aerostat design (approximately 40 ft long by 25 ft in diameter), which lifts the heavy elastic nylon towline. Inflating and launching an aerostat of this magnitude is best done with large safety margins using a separate floating platform or buoy away from the launching ship. Its tether is picked up by the supply ship afterwards.

A smaller aerostat design is preferred, launched from the supply ship. This aerostat lifts a feeder line for the intercept loop. The feeder line is either connected to the elastic nylon towline on the deck or directly to the logistics glider if the elastic section is supplied by the tug. A light, strong feeder loop with standoff station poles is the aerostat's primary payload, although a sensor suite is still recommended. A sub-30-minute cycle time to lower, feed, and loft an aerostat back to altitude can be expected.

Wind conditions and ship rigging affect the safe assembly and launch of aerostats from ships. Ideally the aerostat design size and assembly procedures would allow for the safe launch from the flight deck of the supply ship.

There are considerable design challenges.
- Assembly of an aerostat from the deck of any supply ship
- Lifting the weight of the feeder section of towline
- Flight procedures and glider snatch angles
- Selecting a strong and lightweight feeder section of towline
- Mounting station poles above the aerostat
- Feeding and refeeding of the towline onto the station poles
- Stability control in winds and seas up through sea state 5
- Automated collision avoidance for too-low tug
- Deck winch and control
- Communications by the sensor suite

2.5.5.9 Flight Preparation.
Final preflight assembly and safety checks are required, as well as potentially tasking the flight computer for autonomous operation.

3   ENGINEERING MODELS

Systems engineering defines the "what" of an object (in this case, systems of subsystems) and its interactions.   Operations research determines the "why" of decision-making. Modeling and simulation are the tools used in "how" these disciplines represent the reality.

Modeling uses the salient characteristics of an object for an abstract representation. The previous chapter builds a conceptual model from the problem space toward a common understanding of iHL-relevant values in perspective. This chapter describes the historical data in physics terminology for algorithmic representations of the conceptual snatch pickup model. This modeling approach supports tradeoff decision-making and, with wise planning, model validation and reuse during the iHL systems' entire lifecycle: design, production, asset management, flight control, navigation, etc. The best approach for model reuse is to algorithmically represent those conceptual aspects in a computational format.

The iHL delivery concept is based upon previously operational systems, integrated together with modern systems engineering. The historically proven systems are modeled together as the baseline system with the least risk to duplicate and prove iHL is viable.

The critical new requirement is the helipad launch, with the following differences to prior operational experiences.
- Shorter launch distances when at-sea
- Options for higher Gross Vehicle Weights
- Higher towline reliability
- Unoccupied launches

Helipad-based glider snatch at sea is modeled in incremental changes from the baseline model to constrain programmatic risk during development.

3.1 Historical Glider Snatch

The historical record does not provide exact performance data for any takeoff distances in the range of helipad snatch. There are anecdotal experiences of empty CG-4A takeoffs in this range,[19] but nothing is officially documented. There is very little surviving measurement data available to accurately baseline snatch pickup physics, so projections are based upon one documented measurement and several films. The most basic form of modeling has always been by presenting a story as a representation of the reality. The following examples describe the model of cargo glider snatch operations.

History has been a harsh judge on WWII glider acquisition oversight, contracting quality, designed intent, and operational effectiveness. There is much emotion associated with

58

WWII gliders that needs to be separated from the technical assessments in this report and a baseline representation of iHL. By today's standards the technological novelties and primitive vertical insertion tactics, techniques, and procedures (TTP) of WWII cargo gliders are moot. Systems engineering had yet to be invented and even the modern processes of military supply chains and lessons learned were in their infancy.

3.1.1 Anecdotal Scenarios

The general nature and milestones of cargo gliders and glider snatch have been described, forming a model of the historical system as designed. Now some speculative historical situations are analyzed for a greater understanding of WWII era gliders and snatch. To explore the scope, outlying conditions are described. Some being negative, they explore unusual conditions or unexpected problems to gain insight into the expected performance.

3.1.1.1 Public Perception.

Snatch pickup at Clinton County Army Air Field suffered only towline breaks without injury. Gliders and glider snatch was a classified program, at least until D-Day, so there was limited public exposure. This exposure was typically air exhibition demonstrations at glider factories and later at air shows for bond drives.[10]

At an August 1943 St. Louis demonstration, a chilling photograph of a dismembered CG-4A was taken just moments before impact.[10] It likely appeared in newspapers around the nation and remained in the public's consciousness. Then reports of spectacular losses from invasion glider operations began with D-Day. That crash's image probably was the only visual reference for many. This was despite a quality assurance process instituted after the St. Louis crash, which prevented any repeats of wing spar failure. Any concern over the sturdiness of the CG-4A is baseless.[10] However, that image may have manifested into a generation's unaddressed fear of engineless flight as invasion gliders faded immediately after WWII.

This is the negative side of two emotional extremes associated with gliders in general. It is a common misperception that an engine implies safety to escape trouble while engineless flight somehow entails greater complexity, skill, or risk. Then there is the positive allure of graceful, efficient, silent, non-polluting, bird-like flight. Both of these perceptions will be encountered throughout the lifetime of iHL.

3.1.1.2 Greenland Rescue Analysis.

No Air Force accident report was filed on the two wrecked Greenland rescue gliders, whereas the two airplanes' accident reports were quite detailed. In fact the message traffic described the gliders only by model "CG-15A" or just "GLIDER" rather than assigned serial number or unique call sign. So the official cause remains unknown.

These gliders were specifically stripped of unnecessary weight for this operation.[21] Now for an operationally proven glider snatch system to fail once could be attributed to a chance accident or piloting error. But these were experienced veterans with arctic rescue training, so for a proven capability to fail repeatedly on different missions is potentially a systemic shortfall. The following speculative causes provide insight into the factors to consider with towline failure in general:

- It is likely that the thin air at such a high elevation decreased the glider's takeoff performance, requiring a longer distance to match the tug's speed. This increased the separation between glider and tug beyond the nylon towline's elastic capacity. It is unclear if a longer towline may have survived the increased separation distance.

- Similarly the thin air increased the C-54's stall speed above the top end for elastic separation, or more plausibly it decreased its overall performance in the climbing maneuver that slows separation while accelerating the glider.[15] The Shangri-La rescue was "unusual" in that it was performed at 5,000 ft elevation[10] while Greenland was at 8,000 ft.

- The gliders were the heavier CG-15A model with shorter wings than the CG4A, indicating a higher rotation speed. There were first 12 and then 13 men onboard, on the high side of payload capacity. Greater weight increases rotation speed above the empty specified weight.

- A C-54 with model 160 snatch gear was en route from Wright Patterson after the second failure;[21] however, the employed rescue configurations were not listed. The C-54 tug was usually equipped with thicker winch cabling and the larger winch model. Some CG-15A gliders were equipped with heaters for arctic use. The gliders were the heavier CG-15A model, thus significantly increasing the towline strain during the snatch maneuver.[15]

- "Chug-chug" happens as the towline tension rises quickly and the load over-accelerates, causing a slacking, then the load is reapplied, etc. Longitudinal tension waves are generated. When these waves meet a mass or end point, there is a (theoretical) doubling of tension at that point.[13]

3.1.2 Baseline Physics Models

A baseline of cargo glider specifications builds a baseline of models toward the helipad snatch model. Models include the towline, expected glider payload ranges, and kinematics during launch.

3.1.2.1 Towline.

Most available snatch pickup towline data is based around the CG-4A. An unused nylon towline would stretch approximately 45% of its 225-ft length during its first snatch pickup, then 30% – 35% a second time, and about 25% for all succeeding snatches.[19] There was 1,000 to 1,100 ft of 5/8-in. steel cable on the winch.[38] The payout of the steel cable was usually about 600 ft[23] and the average G force on the glider was 7/10ths of a G for up to 6-1/2 seconds.[40] A C-47 would intercept the loop at a predetermined speed between 120 to 145 mph and slow to 95 to 105 mph depending on the weight of the glider.[11] When the glider reached the speed of the tug, the winch would automatically brake (if properly set by an experienced winch operator) and the steel cable would be reeled back onto the winch drum bringing the nylon line and pickup hook up to the bottom of the door of the C-47.[15]

There is a comment that a CG-13A snatch used a thicker towline.[10] An official discussion states "A 1-3/8–in. pickup rope was used with a 15/16–in. safety link"[29] for one CG-10 snatch. It is presumed that these heavier gliders required a thicker nylon section for the higher forces, and it may have helped the reliability when snatching the other models when heavily loaded. It is presumed the thicker towline was to be issued along with the heavier snatch equipments in conjunction with the heavier gliders, although no documentation explicitly notes this transition actually taking place for the heavy gliders that were fielded.

The recommended tug for the CG-10A was the more powerful C-46 rather than the commonly used C-47.[29] Documented snatch pickup testing of the heavier gliders often used a B-17F. Later the C-54 became the preferred tug.

3.1.2.2 Payload Model.

The useful load capacity of several cargo glider models is listed in Table 15. The glider's empty weight is subtracted from both the maximum specified and operationally tested weights to show payload range. This range is then listed as a percentage of GVW. These payload or useful load models do not account for packaging and other transported equipments not directly consumed by the warfighter. For comparison, data is shown for one C-47 converted into an experimental glider,[10] and hypothetical conversion data from Table 22, "C-130T Glider Conversion Weights." For modeling purposes it is assumed that the C-130's regular payload capacity still applies.

Table 15. Glider Payload Ranges

Glider	GVW Range (lbs)	Payload Range (lbs)	Useful Load (% of GVW)
CG-4A	3,500 - 9,000	4,000 - 5,500	53 - 61
CG-10A	23,000 - 32,000	11,000 - 20,000	49 - 62
CG-13A	8,900 - 18,900	10,000	53
XCG-17	10,000 - 25,000	15,000	60
C-130T conversion	48,600 - 175,600	57,600 - 63,700	33 - 54
CH-53E	33,226 - 73,500	36,000	52 - 54
MV-22	33,140 - 60,500	19,460	37 - 45

The two rotorcraft models are provided to compare sea-based transport's useful load percentages to the glider approaches. Conversely this implies the significant percentage of non-consumable weight at launch and its requisite energies spent in delivering to the LZ. This implication concerns the amount of fuel and its associated processing footprint that the Sea base must apply to weights delivered to the LZ and not consumed by the warfighter.

It is interesting that the upper end of useful glider payload percentage is around 60% despite the pedigree of these cargo designs. Payloads of 60% GVW will be used to model logistics glider concepts presented in this report. It is anticipated that a state-of-the-art logistics glider designed specifically for the Sea base can make even 60% a conservative specification.

3.1.2.3 Distance and Acceleration.

Measured data for WWII glider snatch is very rare, especially for the distance on the ground either taking off or landing. The interest was wholly toward clearing an obstacle while airborne when landing or snatching. Several videos allow limited kinematic estimation by time as measured by frame count. Flight Test Report TSFTE-1976 plots CG-10A runway-based takeoff distances, but not for snatch pickup.

The happenstance of a towline break during the XCG-10's second operational test flight provides the only documented short field snatch measurement.[29] Assuming it landed at the specified 62 mph loaded stall speed, its measured landing and takeoff distances on the field are listed in Table 16, with the remaining data based upon kinematics.

Table 16. XCG-10 Short Field Snatch

CG-10 in Field (using flaps)	Field (ft)	Time (sec)	Tug Flight (ft)	Acceleration (G)
Land at 23,750 lbs	320	7.04	-	-0.40
Snatch at 13,350 lbs	0	0.00	0	0.00
Rotate at 50 mph	120	3.27	390	0.70
v_{B-17} of 160 mph	-	10.48	1,230	0.70

The 1946 film[39] at the Greenville, South Carolina airport shows routine glider snatch operations. The B-17 was recently upgraded with an energy-absorbing winch and the engineers were experimenting with some aspect of the system. The tug performs a preparatory flyby above the station poles for several reasons. During experimentation, radio contact was required prior to pickup and was not always reliable. During air shows it prepared the audience for the snatch maneuver. Usually they "just did it."[11]

The next two tables estimate the cumulative kinematics of two separate glider models snatched in the film. The runway is considered level with the least possible surface friction during the takeoff roll. The pilots are not attempting a minimum distance takeoff as might occur in the field. Flaps are not shown in use in any of the runway takeoffs. It is remembered that those gliders had a sandbag payload and it was a warm late-spring day.[11] The estimated data presented in these tables are based upon independent body, constant acceleration kinematics by counting frames of the translated 8mm film in a DVD player. The estimated distance to roll the length of the glider is based upon its shadow on the pavement. To show an uncertainty in measurement, two liftoff points are identified as each wheel leaves the ground. The final wheel liftoff acceleration is repeated for further estimating the end of the snatch pickup maneuver.

The continuous, rear view sequence of a CG-4A snatch is used for the time and runway distance in Table 17. The CG-4A demonstrates its tendency to slam the nose down upon towline tensioning, likely due to its high, off-axis hookup location. The B-17 intercept velocity is remembered to be 120 mph.[11]

Table 17. 1946 Runway Snatch of CG-4A

CG-4 runway film	Time (sec)	Tug Flight (ft)	Runway (ft)	Acceleration (G)
B-17 start frame	0.00	0	-	-
B-17 contact station	0.67	60	-	-
CG-4 tail movement	1.30	110	0	0.00
Roll length of shadow	3.30	290	48	0.75
Lift right wheel at 50 mph	4.17	370	105	0.80
Lift left wheel at 50 mph	4.40	390	114	0.74
v_{B-17} of 120 mph	6.37	560	-	0.74

CG-10A snatch calculations are shown in Table 18. The B-17 intercept velocity is remembered to be 130 mph. The YCG-10A model in the film is remembered to gross in excess of 20,000 lbs, if not 25,000 lbs.[11] Its speed at flight rotation is based on documented observation.[28]

Table 18. 1946 Runway Snatch of YCG-10A

CG-10 runway film (no flaps)	Time (sec)	Tug Flight (ft)	Runway (ft)	Acceleration (G)
B-17 low point	0.00	0	-	-
B-17 contact station	0.75	70	-	-
CG-10 rolls forward	3.57	340	0	0.00
Roll length of shadow	7.10	680	70	0.35
Lift right wheel at 85 mph	9.33	890	360	0.67
Lift left wheel at 85 mph	9.63	920	380	0.64
v_{B-17} of 130 mph	11.35	1080	-	0.64

WWII glider snatch acceleration ranges about the documented 0.7 G.[40]

3.2 Physics of Helipad Snatch

Two critical logistics glider performance criteria during helipad snatch are GVW and its launch acceleration. Equation 1 ($F=ma$) applies this required force to the logistics glider for successful launch. The greater the payload percentage and the lower the acceleration, the easier it is to design helipad snatch. Acceleration is determined by the following:
- Logistics glider airfoil design, as affecting its rotation speed
- Logistics glider wheel print
- Helipad width for the available takeoff distance

The useable weight or payload percentage is the logistics glider's design measure of useful mass as a percentage of GVW. The remaining percentage is considered necessary operating overhead and should be minimized.

Additional measures influence the applied force but are not explored in this report. Determining the practical ranges of these factors is recommended in a follow-on effort.

- Helipad motion in sea state
- Crosswinds
- Towline geometry
- Tug phugoid - The tug during the snatch maneuver follows a phugoid or arc near the bottom of a circle or ellipse: pitching up after towline intercept and then leveling into tow.
- Tug performance specifications

The model of helipad glider snatch uses a helipad width of 32 meters. The distance to achieve flight rotation speed is restricted to this distance minus the subject glider's wheel print. The wheel print is the leading edge of the front wheel to the rear edge of the rear wheels. This distance along with its modeled flight rotation speed is incorporated into Table 19 to show the time and acceleration necessary for successful helipad snatch. The XG-33 logistics glider is modeled with a higher rotation speed due to its greater payload capacity. The baseline XG-22 logistics glider's acceleration is then used to compute the time and distance at which it reaches the speed of 139 knots (160 mph).

Table 19. Helipad Snatch Model

Helipad Glider	Distance (ft)	Velocity (knots)	Time (sec)	Acceleration (G)
Maximum Flaps	0.00	0.0	0.00	0.00
CG-10A rotation	76.9	43.4	2.10	1.09
XG-21 Logistics Glider	90.0	43.4	2.43	0.93
XG-22 Logistics Glider	89.2	43.4	2.44	0.94
XG-33 Strap-on Glider	85.0	53.9	1.87	1.51
v_{V-22} of 139 knots	914	139	5.31	0.94

The simplest free body diagram of the glider during launch has the sum of all forces acting on the mass of the glide as a horizontal acceleration vector. Estimating from the few data points of the CG-4A and CG-10A, the first three models in Table 20 calculates a range of resultant forces experienced during WWII.

Table 20. Snatch Force on Glider

Snatch Model	Distance (ft)	Acceleration (G)	Mass (lbs)	Force (kN)
CG-4A runway	114	0.74	4,000 - 7,500	13 - 25
CG-10A runway	380	0.64	18,000 - 25,000	51 - 71
CG-10A field	120	0.70	13,350	41
CG-10A helipad	77	1.09	18,000 - 25,000	87 - 120
XG-21 Logistics Glider	90	0.93	5,000 - 11,300	20 - 46
XG-22 Logistics Glider	89	0.94	9,900 - 22,600	41 - 94
XG-33 Strap-on Glider	85	1.51	24,200 - 42,000	160 - 280
Amphibious Log Glider	201	0.64	36,300 - 65,100	100 - 185

The CG-10A is then modeled as being snatched from a helipad as a baseline comparison. The mass stays the same, but the shortened distance significantly increases the force required. The three variants of logistics gliders are likewise launched from the helipad at 55.6 knots with each GVW ranging with a typical payload from Table 8, "All-Aerial MEB Sustainment," up to maximum design weight. The amphibious logistics glider is not snatched from a helipad as are the other logistics gliders; rather it is snatched from the water surface. The loaded amphibious logistics glider's mass is arbitrarily accelerated at the CG-10A runway rate for the displayed distance and force.

The requisite XG-21 logistics glider forces are completely within WWII achievements. The XG-22 logistics glider force range starts at the measured WWII CG-10A short field snatch pickup. Fully loaded, the XG-22 could require more than double that force. The fully loaded XG-33 might require up to three or four times WWII forces mostly due to its heavy weight but also due to the slightly longer wheelbase. The amphibious glider force range appears between these last two, but has room to adjust as needed.

The increased force needed to perform helipad snatch at maximum GVW can come from any combination of the following sources:
- A higher tug velocity providing greater momentum transfer. WWII intercept speed was limited by the boom slicing the nylon loop, whereas iHL can use a stronger material for the feeder line. The CH-53 does not have the kind of stall speed restriction as fixed wing tugs.
- A greater tug engine thrust to accelerate the two-vehicle system. The MV-22 has approximately 20% greater horsepower than the B-17 for approximately the same mass.
- An improvement in towline geometry. The horizontal pull of the towline is the cosine of its angle to the glider. The ship deck and aerostat are higher than ground-based intercept, allowing for new geometries. There are also different towline lengths possible with today's materials compared to WWII equipment.
- An improvement to the tug phugoid. Physics modeling of the tug maneuver between intercept and glider liftoff may provide additional force. Each tug model may have maneuver path limitations with a taunt towline, or stress limits to its airframe and crew.
- Reduce the logistics glider mass at helipad snatch. One way uses the typical payload model in Table 8, "All-Aerial MEB Sustainment." Maximum payload takeoffs are then only available to the less restricted ground and runway launch weights in Table 5, "Logistics Glider Size and Payloads."
- As listed in the beginning of this section, design the logistics glider for minimal requisite forces during helipad snatch.
  - Another way to reduce its mass is to design a greater maximum payload percentage than the modeled 60%.
  - Design the airfoil for lower rotation speed while maintaining or sacrificing high redline speed.
  - Novel footprint design.

It is proposed that the logistics glider's wheels have a counter launch force until that instant in which sufficient force is present on the towline to launch in the width of a helipad. This can be accomplished by braking or otherwise by a wheel design providing sufficient inertia to delay movement across the helipad.

It is important to further account for the physical forces in helipad snatch. Per Equation 1 force has significant bearing on the allowed mass or GVW of the logistics glider. This has significant influence upon the decisions trading off iHL capabilities. For example, the acceleration across the helipad has a limited trade space with the design choices for a logistics glider's wheel print. In contrast the trade space regarding logistics glider weight is large. Accounting for all of the forces present at helipad snatch will indicate the range of logistics glider weights that can be snatched. If this is not an issue, then many logistics glider design alternatives can be explored and the required automation for its load and launch cycle is minimal. But if launch weight is a limiting factor across the iHL system, then logistics glider design shall emphasize a high useable payload percentage and minimal overhead weight. Likewise this influences the need for advanced automation to meet the Sea base load and launch cycle.

3.3 Resupply Modeling Process

The basis in representing Year 2015 MEB resupply using iHL in this report is the following modeling sequence:
1. Integer multiples of the TEU footprint volume derive maximum shapes for the logistics glider in Tables 6 and 7, "Logistics Glider Assemblies."
2. The wing and lifting body surface area of those shapes determine the lifting force for the GVW of the logistics glider in Table 13, "Airfoil Lift Capacity."
3. A cargo payload model of 60% of GVW is based upon the findings of Table 15, "Glider Payload Ranges."
4. Official ashore MEB consumption quantities and MV-22 sortie densities for typical consumable resupply materiel are averaged for a count of logistics glider launches using a generous 50% payload density in Table 8, "All-Aerial MEB Sustainment."
5. The daily number of launches is divided among the available helipads in an arbitrary 8-hour day to derive the helipad launch cycle time also in Table 8.
6. A round trip delivery scenario from Table 14, "Towed Delivery Scenarios," provides a worst-case tug orbit time. This derives the number of sorties per 8-hour day for a tug. The daily number of launches represents single tow sorties, which are divided by the sortie count to get the number of tugs required in Table 9, "Single Shift Tug Count." The single tow sortie count is reduced by a factor of 2 for double and 3 for triple snatch pickup sorties for the number of tugs needed.
7. The retrograde sortie model simply assumes that a repeat of the delivery missions occurs at a later time. There is no interleaving for greater efficiency yet. The dual tow of the XG-24 built from two XG-22's halves this number of sorties.

This is not necessarily a valid scenario, only a quick-look at the numbers involved.

For modeling logistics glider launch physics:
1. The helipad width minus the logistics glider wheel print is the distance available to accelerate in Table 20, "Snatch Force on Glider."
2. This is distance and the mass as GVW from Table 13, "Airfoil Lift Capacity," determines the resultant force upon the logistics glider during helipad launch.

4 DEVELOPMENT PLAN

The management aphorism "fast, good, or cheap: pick any two" is but a generality. This chapter identifies the critical components toward demonstrating a prototype iHL concept.

From its name "interim," there is a limit on the return on investment, so the sooner iHL goes into operation, the better the value in developing it. Any interim capability with a conservative discovery & invention process or slipping development schedule reduces that return on investment. For interim heavy airlift, slipping the schedule too far to the right eventually eliminates any return, such as the point when a tactical Sea-based Joint Heavy Lift concept is available.

A fast development schedule necessitates aggressive risk reduction. Each of the three developmental phases described in this chapter measure an expected level of performance before completing the next phase. While they may be combined or overlapped to eliminate, say the middle phase's hardware, conceptually these phases manage developmental risks. Risk is valued in the change from a known baseline, so a baseline operational model is first established and variation from that baseline is tracked.

Performance drivers are the key to any operational acceptance.
- The logistics glider must become airborne from a supply ship.
- Snatch dynamics must be safe and reliable in operating environments.
- PHST of both cargo and logistics glider must be effective when not flying.

While the aviation and materials performance challenges are not trivial, WWII technological challenges are completely realistic with modern technology and engineering standards. However, the necessary PHST, automation, and TTP to make this performance-sensitive system realistic within the Sea base supply chain are not trivial and have yet to be proven.

While potentially counterintuitive, the logistics glider air vehicle development needs to be driven by the logistics community. iHL is served by the aviation community during one leg of its tactical lifecycle, but is a key component of the surface and ground supply organizations for all others. Ownership of these associations needs recognition. Surface and ground interests must be represented programmatically to the respective aviation, supply, and distribution agencies performing iHL development. Difficult system tradeoffs will be made by this diverse community. Cross-discipline modeling and simulation describes and integrates these interfaces by valuing the acquisition metrics of iHL: unit cost, schedule, and performance.

The WWII juxtaposition between unit production cost and operational expectation is a likely danger to repeat. It is highly unlikely, especially initially, that the price performance of the logistics glider will achieve a disposable unit cost. The vehicle is many times larger than modern two-seat sailplanes. The threshold between patching logistics gliders for reuse and their disposal will be difficult to realistically model for decision tradeoffs.

The initial performance modeling of the iHL concept is software-based and targets critical subcomponents for perspective and relevance, and then integrates those models into an end to-end physics-based modeling and simulation. Model reuse is a factor throughout the system's lifecycle. The purpose of this phase is to determine the scope and priority of key performance parameters and conditions, such as expected maximum payload weight and volume upon helipad launch.

At the same time, teaming to leverage the strengths of Navy labs is necessary to establish a community knowledge base, component demonstrations, and plans for spiral insertion of new science and technologies, and customer education and involvement.

Just mimicking WWII techniques does not guarantee launch in helipad width at sea state. An iHL baseline is established and then carefully changed. There is insufficient data from historical records to build acceptable models so modern experimental data must eventually be gathered. A full-scale, proof-of-concept demonstration could conceivably be built by just copying CG-10A and STARS blueprints and launching from a helipad. Rather than a physical demonstration, this is conceptually performed in software models and simulation. This becomes an initial baseline to start modeling each change to that baseline to manage new risks.

Novel technologies and techniques are a higher risk since they have no operational experience base. The proven aspects of iHL not only require rediscovery but also have not been previously integrated. The following components are key areas needing an increased capability over any previous experience.
- Towline, winch, boom, and intercept performance are at or above WWII experiences.
- Air vehicle processed as ISO-standard cargo.
- Sea base automation and ground forces interacting with logistics gliders: loading, unloading, and performing flight preparation in a reliable processing cycle.
- Aerostat or balloon intercept by a Sea base tug.
- Retrograde delivery of a fixed-wing vehicle to the Sea base.
- Autonomous vehicle operation using Open Architecture precepts.

Table 21 lists the critical path items with their insertion into developmental demonstrations. These insertion points indicate the latest phase for which a capability is required; sooner is desirable for risk and acquisition cost reduction. The proof-of-concept demonstration requires minimal technologies since many components have operated in the past. Most new technologies are spirally inserted somewhat randomly in the middle phase as they mature. New capabilities are of course required before any consideration for transition: these are integrated in the "Technology Spiral Insertion" column and as such are identified as risks to track. Finally, the transition efforts in the last column are required in a prototype demonstration for a program of record.

Those remaining technologies listed in the "Transition Prototype" column of the table may be postponed until actual transition requirements are clearly defined. Their earliest incorporation mitigates risk and costs. Some require maturing the understanding of system cost and performance trade-offs. Given such a diverse community for consensus, they are initially high risk to demonstrate any sooner, but this may change as tradeoff decisions are made.

Table 21. Component Priorities

Sub-Component or Function	Proof of Concept	Technology Spiral Insertion	Transition Prototype
Performance modeling	X	X	X
Management software		X	
Tug boom, winch, etc	X		
V-22 snatch maneuver	X		
CH-53 snatch maneuver		X	
C-130 snatch maneuver		X	
Towline reliability	X		
Towline materials improvement		X	
Towline tension sensor		X	
Autonomous control		X	
Approach and landing sense		X	
Pilot or passengers			X
Sense coop/non-cooperating payload		X	
Airframe	X		X
Construction materials			X
Body Packaging storage transport		X	
Flatbed assembly transport		X	
Daily assembly/disassembly			X
Spine rail(s)		X	
Wheels, brakes, suspension			X
Reusable balloon		X	
Retrograde delivery		X	

First there is concept exploration, verification, and component development. There will be a similar process to WWII experimentation and technology development for increasing the helipad snatch forces and hence GVW launched. This starts as an experimentation phase for data collection using a converted plane simulating the logistics glider. For the fastest development schedule, it is expected that there will be several generations of glider airframes, tug winches, and intercept equipments as prototype discovery migrates to transition prototype demonstration. Eventually a concept prototype is assessed and system models are matured. Finally a transition prototype model will be demonstrated, ideally considered as "revision 0" of the limited rate production units.

All the items in the table need acceptable risk mitigation in place by transition prototype demonstration. A continuous investment in the development of TTP by supply, distribution, and tug crews occurs throughout development. Tug crew training is a prerequisite to any glider snatch experimentation. Initially sea state motion will just modeled.

The efforts described next mitigate high risk areas. Simultaneous experimentation of the medium risk technologies and TTP are integrated as each mature. All the concepts and technologies described influence each other as a system and are integrated into the demonstrations that follow.

It will take full-scale, land-based runway experimentation to derive meaningful data for liftoff distance on the scale of a helipad. All technologies spiral into iHL as they mature. The full-scale items are available for reuse in the next demonstration phase, such as an extra vehicle available for multiple snatch pickup demonstration. Modeling and simulation lowers development costs, unit production costs, and support costs. This keeps iHL fundamentally more cost effective than any similar capabilities for the Sea base.

The initial exploration task delivers this report with the proposition that the iHL concept is viable and warrants further investigation. The next step is the verification of those risky or otherwise new iHL components in the logistics glider's tactical lifecycle. This culminates in a decision to proceed with demonstration. Then the preparatory engineering toward a concept demonstration program begins, starting with ground snatch experimentation.

4.1 Verification Modeling and Simulation

This effort starts with models and then simulates in software those tactical lifecycle components that are novel and hence the highest risk to concept demonstrations in the succeeding phases. This phase concludes its verification studies with the risk mitigation engineering necessary to prepare these models for physical experimentation.

4.1.1 Helipad Snatch

A realistic proof of the physics of helipad snatch at sea is modeled in MATLAB SIMULINK. The additional forces required for helipad snatch over WWII glider snatch are accounted for using modern capabilities. At least one optimal range in tug performance is presented to a tow craft program office for consideration of support for iHL.

The physics modeling includes sea state, cross wind conditions, and rudimentary models of the winch and towline to verify the transfer of forces from the tug to the logistics glider. This is a placeholder until more detailed models are obtained from the respective design studies to follow.

This effort involves publishing, presentations, and review support to the aviation community for logistics glider flight capability and tug selection.

4.1.1.1 Structural Analysis.

An analysis of the structural forces upon the logistics glider determines modern body designs most suitable to helipad launch, ground landing, and retrograde delivery. This effort:

- Assesses the body-on-frame approach and prepares the logistics glider designs to follow.
- Develops the mechanical attachments of the bottom to the fuselage.
- Addresses the economics of logistics glider materials.

4.1.1.2 Simulation Visualization.

There is much historical baggage and distracting connotation associated with the military application of cargo gliders. To help visualize the iHL concept of operations, an end-to-end simulation visualization is the recommended basis for educating and furthering community discussion. The SIMULINK output for helipad snatch is directly converted into a MultiGen-Paradigm, Inc. Vega Prime three-dimensional visualization of interacting OpenFlight format models constructed in MultiGen Creator.

Starting with an AOE class ship in sea state, a video of the tug, aerostat, towline, and a logistics glider interacting will present helipad snatch. Then this simulation is expanded into representative aspects of iHL tactical lifecycle. Finally adding payloads and personnel within the visualization will support community education of the CONOPS. This brings the Sea base aviation, supply, and ground communities into a common understanding of the proposed iHL operation.

4.1.2 Experimental Glider Design

The XG-21 (or XG-31) is designed to verify iHL models as these are the smallest and least-risk logistics gliders. Although it might not single-handedly support the objective of all-aerial MEB resupply, it is the simplest step in that direction. This step provides the following:

- Verification of the model from this report
- Modeling details to all the other iHL efforts

4.1.3 Retrograde Transition Concept

Risk mitigation for the novel retrograde approach for iHL is necessary by exploring, engineering, and eventually demonstrating one or more techniques.

- Task times to sling-carry a fixed-wing vehicle from the LZ to the Sea base.
- A 3-point harness used when stalling a flying logistics glider into a sling-carried delivery to the Sea base.
- Other, more efficient delivery methods to the Sea base.

4.1.4 Basing Footprint Model

Three explorations of the impact and integration of the iHL system upon the expeditionary supply chain are produced. These studies require a preliminary logistics glider design to work from, and take into account external, internal, and ergonomics of iHL processing.

For the incorporation of the logistics glider as a vehicle, iHL system performance, operator interaction, and PHST influence its design as much as the physics of flight. The logistics glider exterior's PHST exploration involves many overlapping aspects of vehicle use. Implications to the exterior design of the vehicle include the following, for example:

- ISO standard lift points
- ISO standard stacking capability
- Tie downs
- Towline operation
- Sling load
- Ground and deck towing

The logistics glider interior's PHST exploration involves many overlapping aspects of vehicle use. Implications to the interior design of the vehicle include the following, for example:

- Automated loading equipments and their interfaces
- Manual unloading operations on the ground or aboard atypical sea platforms
- Tie downs
- Spinal rail system
- Ground vehicle winch interfaces

Logistics glider ergonomic considerations include, for example, the onboard computer and electronics suite. It should be ideally placed within a suitcase-style package for easy access, removal, and replacement. There is no requirement for expensive, high power, large volume electronics, so the emphasis is on user access and maintenance at the operational or intermediate level. Other examples include the following:

- Jack lifts
- Unlocking, such as after unattended landing
- Vehicle assembly and disassembly
- Interior lighting, windows
- Access door and ramp

4.1.4.1 Advanced Base.

Models of iHL—in terms of any USTRANSCOM impacts to Advanced Base storage, processing, maintenance, and the equipments and personnel required—are developed and documented.

4.1.4.2 Sea Base.

Models of iHL—in terms of Navy impacts to Sea base storage, processing, maintenance space, the automation equipments and personnel required, and fuel consumption—are developed and documented.

4.1.4.3 Ground Operations.

Models of iHL—in terms of impacts to Marine air, ground, and supply operations for storing, processing, and maintaining the necessary equipments and personnel—are developed and documented.

4.1.5 Pickup Equipments

Before ground-snatch experimentation occurs, the following engineering efforts are needed.

4.1.5.1 Ground Station.

Any intercept station development must first duplicate the WWII ground station. This establishes a foundation of experience for towline changes, approach angles, and wind conditions. This would be evaluated for ground operations in retrograde or partial load deliveries. These station poles, clip springs, and documentation will apply to all following phases including aerostat design.

4.1.5.2 Aerostat.

Separating the intercept station from sea motion requires analysis, planning, and engineering. The plan is to range the intercept between 500 ft and 999 ft above ground level (AGL) for realistic towline experimentation. This is typically a stable air zone for balloons, although this is less true in the littoral than other regions.

If there are delays in producing a safe, balloon-based intercept, a tower could be built in the meantime to loft the towline loop to this altitude. Alternatively a STARS-type demonstration could be engineered quickly with a disposable balloon technology for an early demonstration of balloon-based intercept. This would have to be weighed against long-term use of a tower or accelerating any other approach.

The preferred approach is to develop a new, reusable aerostat-based system for a stable, safe, and reliable contact station in varying weather conditions. Initially this would be used in ground-based experimentation, then for shipboard operations. Once delivered, all helipad snatch experimentation would use this technique.

Consideration is necessary for the aerostat's launch, recovery, and PHST from supply ships in the littoral for sea states 0 through 5. Simulation of sea state conditions on the aerostat, all lines, and its base during approach and intercept will define the upper range of operations. Deliverables also include flight patterns for tug approach, intercept, abort, hold, retry, etc., for flight controller review.

4.1.5.3 Boom.

The boom attachment to the tug will require at least the following:
- Tug selection
- Modern engineering for potentially higher intercept speeds
- Safety of flight considerations

4.1.5.4 Winch.

Constructing a winch capable of helipad snatch energies is the greatest unknown technology to iHL development. Rediscovery and improvement upon the last pickup winch technology will be required. Significant automation over WWII operation is necessary.

Between multiple snatch pickups, any WWII-style manual transfer of towlines is too dangerous and labor intensive. Multiple towlines involve the line transferring off the winch drum and require automation so that the winch and cable line gets reused for the second and third logistics glider snatches.

The winch should communicate what state it is in. A green light turned on means that a positive connection has been made after the intercept and the towline is connected (but not tensioned yet). When tensioning begins, the light turns yellow. Upon full tension it is red. Off means no towline is connected. There may be more complex states for multiple towlines or tug climb.

There are two communication paths into the winch controller for expected tow weight when tensioning the towline. Asset management information comes electrically from the tug such as expected logistics glider model, summary information from inventory control software, and any broadcast information from the glider.

From the towline comes information from inside the glider, such as serial number and any internally estimated gross weight. There are several seconds upon connection before tensioning that the glider can report to the winch for reconciliation of data for winch tension response.

4.1.6 Towline

The performance envelope window for safe and reliable operation needs to be identified, with higher reliability than previously with glider snatch. The key performance parameters for GVW are the nylon rope tension force and safe elongation performance ranges. There is the added complexity of a lightweight feeder line in the towline system that is carried by the aerostat. The towline forces need to be "sensored" for model data points, both nondestructively and destructively.

4.1.6.1 Nondestructive Engineering.

Three efforts are needed to better model the towline's performance during logistics glider snatch. They can be performed simultaneously or sequentially.
- Research the physics of repeat polyamide cycles in the low hertz range to develop an accurate model of the nature of towline stresses during snatch.

- Obtain limited laboratory measurements to guide field experimentation.
- Conduct snatch experiments with loaded gliders and tugs at speed.

4.1.6.2 Destructive Experimentation.

Determining the towline performance window should not be done with full-scale aircraft due to the damage that a failure in the highly tensioned lines can cause to the experimental logistics glider, personnel, and tug. A test environment should simulate snatch conditions to safely explore failure stresses upon the towline.

4.1.6.3 Towline Sensing.

Valuable control information can come from just sensing the logistics glider's connection to the towline. At a minimum it is necessary to determine whether the towline is connected, under tension, or not. This is communicated to the control system.

Advanced technology development includes monitoring the nature of towline tension, such as direction and force of the tension. This is critical for autonomously recognizing the logistics glider's state of operation. This effort develops the algorithms to determine by the behavior of towline forces the logistic glider's environment such as ground towing, the preflight tow release check, snatch intercept, multiple snatch pickups, release, etc.

4.1.7 Technologies Spiral

The various technologies listed are recommended for maturation and insertion in parallel with the concept demonstration efforts described afterwards.

4.1.7.1 Decision Support Software.

Increasing delivered payload densities significantly greater than the 50% payload density model presented in Table 8, "All-Aerial MEB Sustainment," requires a forecasting and optimization technology with a level of coordination yet unproven with afloat and ashore expeditionary logisticians.

Being an interim system with rapid delivery to the fleet, it is likely there will be various logistics glider classes or capabilities in operation with different performance windows. With dozens to hundreds of vehicles in operation each day, not all can be aboard the Sea base at once. The tracking, coordination, and optimal planning for a family of vehicles are not viable without complex software technologies for Command and Control (C2) integration.

This level of forecast optimization, automated decision support, and sensor integration is technologically feasible, but interest has not been demonstrated to date. The immediate hurdle is merely a requirement for logistics bandwidth.

These technologies are combined into a semi-automated decision support system for the C2 of iHL.

- Asset inventory control and operational tracking
  - o Interfacing to warehousing and asset visibility systems throughout the supply chain
  - o Tracking and asset visibility of tow craft
  - o Automated tracking of logistics gliders on the ground
- Mission planning and course-of-action decision-making
  - o Sea state, meteorology, route planning and replanning
- Load planning and payload management
  - o Warehouse and asset visibility to organize cargo volume, weight, stacking requirements, center of gravity, tie down procedures, proximity specifications, load and unload sequencing, delivery timing issues, etc.
  - o Automated real time sensors to create a launch manifest during loading. Given non-supply ship environments, this is recommended from the inside of the logistics glider. Sensors should detect cooperating payload (e.g., RFID tags) and non-cooperating payload (e.g., weight and trip sensors).

Extending the modeling effort across the entire iHL system to explore the performance envelope will incorporate community suggestions, clarify detail, and quantify alternative concepts into ideal operations and the value of trade space options. These models will be incorporated into the visualization effort above. Formatting simulations into SISO MSDL models (an XML format) targets technology transition into operational asset management and decision-making software tools.

4.1.7.2 Autonomous Operation.

A concept demonstration of helipad snatch could be shown with onboard pilots or by remote pilots. Unattended operation during helipad launch will require an autonomous flight capability with internal decision-making and flight-control technology.

It is recommended that a command, control, and communication software architecture framework be developed and the long-term developmental standards guidance for all autonomous platform profiles, be it combat, surveillance, or logistic. A narrow-minded approach to autonomous glider navigation and guidance would not be interchangeable with combat autonomous development: Glider free flight navigation and LZ targeting is, for example, a subset of search-and-destroy swarm algorithms. While the maximum of three logistics gliders towed per sortie may be debated under the minimum definition of a swarm, the collision avoidance of potentially dozens of logistics gliders arriving at semi-random intervals to a waypoint, into a traffic pattern, onto an unimproved field or parking space is best addressed via onsite autonomous decision-making rather than remote procedural restriction. This is a costly lesson learned from every WWII glider operation.

This problem is not limited to logistics gliders or even air traffic control. Standardizing distributed autonomous system technology development is a proposed NSWCDD concept.

Autonomous flight options while occupied or remotely piloted also need to be explored since this is high risk technology. The capability to switch between autonomous and piloted control needs exploration.

4.1.7.3 One-Way Communications.

This includes all passive navigational guidance technologies received by the logistics glider.

4.1.7.4 Two-Way Communications.

Simple communications may be needed to redirect autonomous operation during free flight, or as an alternative to towline-based communications.

4.1.7.5 Towline Sonic Communication.

There are several seconds after towline intercept and before line tensioning in which the logistics glider and tug can communicate via the towline connecting them. This can help define the safety margins, speeds, and angles for best performing the snatch pickup.

4.1.7.6 Payload Sense.

The automated sensing of the payload helps forecast the physics of logistics glider snatch. To detect cooperating payload (e.g., RFID tags) and non-cooperating payload (e.g., weight and trip sensors) requires new technology integration.

4.1.7.7 Landing.

There are many ground, water, and Sea base landing technologies to be explored. The 3-point transition harness is one new technology recommended for development for iHL.

Landing and taxiing the amphibious logistics glider in water is a concept not otherwise explored in this report.

4.2 Ground Snatch Experiments

Becoming airborne by the edge of the helipad is the definitive logistics glider performance criteria. Experimental data is required to populate the snatch model. Physics modeling will indicate if flight from a helipad is possible, and determines the maximum logistics glider gross vehicle weight. The data needed for finding GVW includes towline elastics and tug climb properties at snatch.

The initial modeling stage can be refined with careful measurements in a series of simulated glider snatch experiments. Scaled experiments are expanded from sized drogue parachutes, then to general aviation gliders on up to the planned military tugs and a stripped cargo plane. The tug may at first be an appropriate test craft before experimentation with operational models.

4.2.1 Data Collection

Considering the vehicles in implementing this effort, this task begins with designing an experiment and procuring sensor recording instrumentation for the following:

- Tow craft position and G-forces
- Towline stress (axial force)
- Positioning of the ends of towline and their acceleration
- Towline elongation

4.2.2 Parachute Snatch

The first snatch rediscovery concept is to simulate the drag of a logistics glider behind a tug using a carefully sized drogue parachute or similar representation of a logistics glider. A parachute simulates the forces of drag and weight of a glider on the tug and towline. The parachute and towline can be snatched initially by recreating the WWII ground station and glider snatch technique.

The experiment then is refined as technologies, safety, and techniques permit to simulate the modern changes to the baseline technique.

- Approach angles for ship helipad winds
- Aerostat or disposable balloon
- Tug climb out maneuver
- Towline improvements

4.2.3 Glider Snatch

The experiment is next expanded to manned sailplanes for small-scale rediscovery of glider snatch, being very mindful of safety margins. Deliverables include revising the WWII documentation for modern naval pilots with anticipated ship helipad angles and cross wind approaches, telemetry data collection for modeling, and a plan for the transition to and the training of military test pilots in all aspects of logistics glider snatch.

4.2.4 Airplane Conversion

The final stage of this phase of experimentation for data collection is a comparable weight simulation of full-scale glider snatch using a military tug. The experiment involves stripping an airplane into a glider of approximately the anticipated weight range.

This full-scale experimental vehicle will be snatched from a runway using military test vehicles flown by test pilots. Actual takeoff distance to rotation speed will not be within helipad dimensions since the stripped plane's airfoil is not truly representative of a logistics glider. However, the weight and forces will be within operational ranges. So the speed the cargo plane reaches upon accelerating the distance of a helipad can be a realistic representation.

Towed flight is inherently demonstrated. Landings can be demonstrated as necessary but will not be realistic without the correct airfoil landing distances. Loading and unloading operations can only be demonstrated as allowed by any cargo bay similarities to the logistics glider. Retrograde operations such as VERTREP sling load or snatch and stall techniques can be performed to demonstrate the return to the helipad.

This phase demonstrates separately the stages of the iHL tactical lifecycle, with end-to-end demonstration in the next phase. Parallel maturing technologies are spirally incorporated as they become available.

The initial idea of stripping a C-130 into a glider is not realistic as the total weight of the glider option shown at the bottom of Table 22 exceeds the GVW of all helipad launched logistic gliders in Table 13, "Airfoil Lift Capacity." The selection of an appropriate test airplane will involve trade-offs in cost, performance, and demonstration value.

Table 22. C-130T Glider Conversion Weights

Component	Current (lbs)	Minimal Tug (lbs)	Glider (lbs)	Comments
Basic Structure	42,140	42,140	38,410	Removes cargo ramp
Propulsion Group	17,080	17,080	600	Keep tank & plumbing for APU only
Flight Controls	1,540	1,540	1,540	
Auxiliary Power Unit	700	700	800	
Instruments	610	610	600	
Hydraulic & Pneumatic	1,000	1,000	520	
Electrical	2,450	2,450	1,660	
Avionics	2,550	2,550	1,850	
Furnishings & Equip	4,260	4,260	2,440	
Air Conditioning	1,770	1,770	0	Removed
Anti-Icing	800	800	0	Limits altitude
Load & Handling	700	700	195	Remove cargo tie downs
Ballast	40	40	0	
Payload/Winch/Hook	42,580	800	30	
Weight Empty	117,380	75,600	48,615	
Unusable Fuel & Oil	2,640	2,640	0	
Fuel Weight	54,980	2,260	0	
Total Weight	175,000	80,500	48,615	

4.3 Helipad Snatch Demonstration

This middle phase applies the first of a series of custom built logistics gliders to start concept experimentation and demonstration. At this point operational tug models have been selected and are being modified for demonstration at the end of this phase.

4.3.1 Logistics Glider Construction

It is recommended to target prototype development with the smallest logistics glider concept first and work up in weight and complexity toward the more sophisticated designs. This begins with the mono wing XG-21, followed by the XG-22, then the bi-wing, multiple fuselages, strap-on or alternatively any amphibious variants. Disposable economics are explored after the iHL end-to-end system is verified as technically viable.

4.3.2 Proof of Concept

The key performance criterion is demonstrated with a logistics glider being snatched on the ground within the width of a helipad.

Later this should be performed at sea. An interim demonstration could be the floating of the tethered base from water or simulated sea state and the launch of the logistics glider from shore, land, or barge.

4.3.3 Multiple Snatch Sortie

Including the experimental glider from the data collection phase, full-scale, dual-snatch operations are demonstrated in one sortie, including loading and the launch preparation cycles.

4.4 iHL Prototype Operation

The iHL prototype demonstration exercises all tactical stages of the iHL system in preparation for transition to an operational production and fielding program. This stage converts a ship into an iHL system platform and demonstrates operational scenarios.

The two key iHL parameters are demonstrated.
- Multiple logistics gliders loaded and snatched in each sortie by an operational tug model from a helipad in sea state 5 at night for 8 hours averaging a launch cycle for example, of every 30 minutes.
- Demonstrate likely production vehicle configurations and technologies.

4.4.1 Ship Conversion

The developments toward all necessary modifications and certifications of a Sea base demonstration ship are likely a long-lead item and may start in conjunction with the middle phase. This mitigates risk of new technologies inserted into the Sea base. The selected ship receives the following iHL components:
- Handling equipments
- Automated warehousing
- Assembly facilities
- Logistics glider loading equipments
- iHL helipad operations to NAVAIR standards
- Logistics glider refurbishment and processing facilities

4.5 *Transition*

Some suggested performance characteristics for transition into a production and fielding program of record are described by the following:

- A payload range given external conditions (wind, tug, takeoff distance)
- One to three logistics gliders towed per sortie
- Delivery in less than four minutes after release at 5,000 feet (GPADS)
- Redline speed in excess of 150 knots
- Soft landing: vertical velocity less than 30 feet per second (GPADS)
- Targeting: 50% Circle of Error Probability (CEP) less than 20 m (ERGM)
- Access to and removal of payload using common infantry and CSS equipment
- Daily field assembly and disassembly by Sailors up to sea state 5 and Marines in the field
- Snatch pickup from ground elevations up to 8,000 feet with appropriately scaled payloads, and on-ship helipads in sea states 0 – 5 with expected payloads
- Typical flight preparation in less than 30 minutes

5 SUMMARY RECOMMENDATIONS

This report shows that the historically proven concepts of cargo gliders, snatch pickup, and balloon intercept can be combined into a viable application for Sea base helipad snatch pickup in all-aerial expeditionary resupply. There is a need for rediscovering, relearning, and reapplying these concepts with modern technology and tactics. This starts with recognizing and organizing to connect a diverse iHL system community.

5.1 Ownership

The greatest risk to this effort is in narrowly categorizing iHL as just an air vehicle to build for the supply community to then integrate. iHL is served by the aviation community during only one, albeit critical, stage of tactical operation, but connects sea-based surface and ground supply organizations during all others. Ownership of this connection must be recognized. Those surface and ground interests must be represented programmatically to the respective aviation, supply, and distribution agencies involved in the development and performance of iHL.

Difficult system tradeoffs will be made by this community. The acceptance of Sea base automation, of standardized containerization, of new equipments and TTP in all-air resupply is necessary by all communities. Cross-discipline modeling and simulation describes and integrates these interfaces by valuing the acquisition metrics of iHL: unit and system cost, schedule, and performance.

5.2 Schedule

Being an interim acquisition solution, the development time to initial operating concept drives the return on developmental investment. iHL could become a discovery and invention effort with a "too long" transition schedule. Rather, a faster track of developmental, DoD 5000-based systems engineering is recommended using modeling and simulation to baseline capabilities and interfaces, support tradeoff decision-making, and communicate technically between diverse iHL surface, air, and ground communities.

5.3 Performance

The iHL baseline starts with previously operationally proven components, and then is described by the specific changes and performance improvements using modern technology. This mitigates risk to within those changes. Further modeling and simulation includes end-to-end, physics-based representation of iHL systems and their interfaces to understand and direct developmental decision-making, tradeoffs, and technical direction. These models continue to build launch-on-the-move flight models and throughput performance simulations as new technologies spiral into the demonstration system. The final developmental models are used for transition acquisition specification, training, and production acceptance.

The physical forces during helipad snatch are within WWII operational capability, and yet can grow to significantly exceed it. Section 3.2, "Physics of Helipad Snatch," presents the necessary forces and their impacts to helipad launch. There is a large trade space dependent upon logistics glider weight and its payload as a percentage of GVW. If vehicle weight is not an issue, then many logistics glider design alternatives are available with minimal reliance upon its load and launch cycle. But if launch weight is a limiting factor across the iHL system, then design for a high payload percentage and minimal overhead weight is important. Likewise this influences the Sea base load and launch cycle with advanced automation.

The unit price performance of a logistics glider will be a contentious issue. It is unlikely that it can or even should be produced to disposable metrics. The storage space aboard the Sea base to provide daily connecting vehicles may dictate reuse regardless of unit production cost. The term "low-cost" is often misconstrued in military context but is the likely range for logistics glider production.

There are iHL configurations that can be implemented inefficiently. The concept of control of performance of the Sea base supply chain is necessary. There can be dozens if not hundreds of vehicles involved each day in MEB resupply operations. One transitionable M&S effort into operation is the optimization software and sensor modeling necessary for an automated decision support and forecasting system for asset command and control.

The novel aspects of iHL will have the greatest risk in technology development. These are the following:
- Intercept and retrograde technologies and procedures
- Supply chain performance and its control
- Open Architecture decision-making of autonomous platforms

5.4 Demonstration Phases

The three demonstration phases summarized below may be combined or overlapped. The investment in the development of TTP and expertise by supply, distribution, and tug crews is mandatory throughout this development.
1. Gather sufficient experimentation data to populate models of the iHL system for go/ no-go acquisition decisions. Key technology areas with show-stopping risk include the following:
    a. Launch within helipad dimensions
    b. Towline reliability
    c. Transport, storage, and assembly footprint required

2. Demonstrate the conceptual system operating; first by computer modeling, then scaling up to full-scale simulation. This must show system integration for the following:
   a. Transport
   b. Storage
   c. Assembly
   d. Loading
   e. Flight
   f. Unloading and distribution
   g. Retrograde operation
   h. Repair and support

3. Conduct prototype demonstration of steady resupply cycles from a helipad in sea state 5, at night, with multiple loaded logistics gliders.

5.5 Conclusion

The interim heavy airlift (iHL) concept is explored as a connector for the littoral Sea base to resupply warfighters ashore. Its historical basis is presented from an expeditionary logistics systems engineering perspective. Variants of new logistics glider vehicles and the tactical lifecycle describe the iHL trade space. The effectiveness of the iHL system lies in its efficient use of limited Sea base resources for the launch of consumable materiel: gliders have a very high percentage of vehicle launch weight dedicated to warfighter consumables. Engines, fuel, and crew take off once for many deliveries.

iHL performance scales to sustain expeditionary forces ashore from small units and Distributed Operations up to a Brigade. Using conservative estimates for iHL performance, Year 2015 Sea Base Maneuver Element MEB daily ashore sustainment requirements can be wholly met in an eight-hour shift by logistics glider variants at half-payload capacity on half of the 12 MPF (F) Squadron's helipad- or flight deck-equipped ships, using fewer rotorcraft or tilt-rotorcraft than otherwise possible. The tow craft may be combinations of Sea base or non-sea-based joint Navy, Air Force, Army, Coast Guard, and Coalition air tow assets, and this combination may fluctuate as needed during the resupply operation.

These first-look iHL models indicate a viable system can be built from the proposed concept, but first there needs to be model verification and refinement of technical and scenario detail. It is recommended that, initially, Section 4.1, "Verification Modeling and Simulation," be implemented to present a mathematically verified helipad snatch pickup concept. Payload weight and system performance has significant influence upon the decisions in the tradeoff space.

The follow-on effort initiates iHL presentation to its diverse developmental and operational community. iHL system development warrants further investigation by all communities involved.

6 REFERENCES

1. Department of the Navy, *Naval Transformation Roadmap*, http://www.navy.mil/navydata/transformation/trans-pg48.html.

2. JMS System Science Corporation, *Sea Base Concepts of Operation and Logistics Technology Applications (Performance Analysis and Investment Strategies)*, Contract NSWCDD N00178-05-M-1218, April 2006.

3. Spencer, Leon B., *WWII USAAF Glider Aerial Retrieval System*, white paper.

4. Lewis, W. David and Trimble, William F., *The Airway to Everywhere A History of All American Aviation, 1937-1953*, University of Pittsburg Press, 1989.

5. "Pickup!! Action in WWII," G. Robert Veazey, *Friends Journal*, Vol. 17, No. 1, Spring 1994.

6. http://www.ww2gp.org/

7. Department of Army Air Force, Training Film TF-1-3399, Glider Snatch Pickup by C-47's.

8. Department of Army Air Force, Manual 51-129-2, *Pilot Training Manual for the C-47 Skytrain Aircraft*, September 1945.

9. http://www.440thtroopcarriergroup.org/tcarticle_glidersquadron.shtml

10. Day, Charles L., *Silent Ones: WWII Invasion Glider Test and Experiment*, 2001.

11. Communications with Lee Jett, 2005-2006.

12. All American Aviation, *Air Pick-Up Handbook*, 1947.

13. Communications with Bob Veazey, 2005-2006.

14. Communication with Lloyd Santmyer, 2006.

15. Communications with Charles Day, 2005-2006.

16. Masters, Charles J., *Glidermen of Neptune: The American D-Day Glider Attack*, 1995.

17. Photo01 CG-13 TroyOH.tif through Photo11 CG-13 TroyOH.tif

18. Van Wagner, R. D., *Any Place, Any Time, Any Where: The 1st Air Commandos in WWII*, Schiffer Publishing Ltd., 1998.

19. Communications with Leon Spencer, 2006.

20. http://www.thedropzone.org/pacific/walters.htm

21. Department of the Air Force Press Release RE: 6700-Ext. 75151, Greenland Ice Cap Rescue files, 15 December 1948.

22. Frisbee, John L., *Valor, Journal of the Air Force Magazine*, Vol. 81, No. 3, March 1998. http://www.afa.org/magazine/valor/0398valor.asp

23. http://www.armyairforces.com/forum/m_70228/tm.htm

24. Communications with Gerald "Bud" Berry, 2006.

25. http://www.specwarnet.com/americas/usaf.htm

26. http://www.airbum.com/articles/ArticleWACOGliderCG-4A.html

27. Miller, Marvin, *Underway Replenishment of Naval Ships*, 1992.

28. http://aeroweb.brooklyn.cuny.edu/specs/laiskauf/xcg-10a.htm

29. Department of Army Air Force, AL-113172-1, *Operational and Tactical Suitability of the XCG-10A Glider*, 12 February 1945.

30. Department of Army Air Force, *AAF Glider Model XCG-10A*, 20 December 1943.

31. Hagley Museum and Library online database, All American Engineering Company Records 1937-1975.

32. http://www.armchairgeneral.com/articles.php?p=2325&page=1

33. Department of Air Force, *Munition Resupply at 30,000 Feet*, Special Operations Technology Online Archives, Volume 4, Issue 3, 12 April 2006, http://www.special-operations-technology.com/article.cfm?DocID=1393

34. McCarthy, VADM Justin D. *Seabasing Logistics* slide 29, NDIA 10th Annual Expeditionary Warfare Conference, October 2005.

35. Stewart, Captain (USN) Jim, *Seabasing Logistics* slide 21, ASNE Joint Sea Basing Conference, March 2006.

36. Kaskin, Jonathan, *Seabasing Logistics CONOPs* slide 14, NDIA 10th Annual Expeditionary Warfare Conference, October 2005.

37. Naval Surface Warfare Center Carderock Division, NSWCCD-98-TR–2004/027, *Assessment of Advanced Logistics Delivery System (ALDS) Launch Systems Concepts*, October 2004.

38. Spencer, Leon B., *Military Glider Tow Ropes,* white paper.

39. *Lee Jett piloting B-17*, CG-4A/CG-13A/CG-10A snatch pickups, Spring 1946.

40. Department of Army Air Force, Manual No. 50-17, *Pilot Training Manual for the CG-4A*, March 1945.

7 BIBLIOGRAPHY

Conway, Carle, *The Joy of Soaring*, The Soaring Society of America, 1969.

Day, Charles L., *Silent Ones: WWII Invasion Glider Test and Experiment*, 2001.

Department of the Navy, *Naval Transformation Roadmap*, 2003.

Gallagher, William E., *The U.S. Navy's WWII Floatwing Assault Gliders*, NSM Historical Journal, Vol. 25, No. 2, 2003.

Lewis, W. David and Trimble, William F., *The Airway to Everywhere A History of All American Aviation, 1937-1953*, University of Pittsburg Press, 1989.

Milgram, Judah et al., *Autonomous Glider Systems for Logistics Delivery*, AUVSI 2003 Unmanned Systems Symposium and Exposition, Baltimore MD, Jul 2003.

Masters, Charles J., *Glidermen of Neptune: The American D-Day Glider Attack*, 1995.

Naval Air Warfare Center Aircraft Division Lakehurst, *Shipboard Aviation Facilities Resume* revision AY, NAEC-ENG-7576, 01 January 2005.

Naval Surface Warfare Center Carderock Division, *Assessment of Advanced Logistics Delivery System (ALDS) Launch Systems Concepts*, NSWCCD-98-TR–2004/027, Oct 2004.

Noetzel, Lieutenant Colonel (USAF) Jonathan C., *To War on Tubing and Canvas: A Case Study in the Interrelationships Between Technology, Training, Doctrine and Organization*, Masters Thesis, Air University, Maxwell AFB, May 1992.

Van Wagner, R.D., *Any Place, Any Time, Any Where: The 1st Air Commandos in WWII*, Schiffer Publishing Ltd., 1998.

GERALD "BUD" BERRY COLLECTION

Glider Pickup.doc – Description with two-photo sequence at Wright Field.

Electronic Images

Glider Pickup.jpg – First Normandy Pickup: Bud Berry is the tow pilot with photo taken by Yves Tariel of Paris, France.

PickupN0 1.jpg to PickupN0 3.jpg – "PICK-UP," Volume 5, Number 7, July 1945.

PickupNo 4.jpg – Buffalo Evening News, 30 March 1945, with another local clipping.

CHARLES L. DAY COLLECTION

Printed Articles

The XCG-17 Glider

Development of an Autopilot for WWII Gliders

Electronic Images

CG4Awatertank_HCmod.jpg – rear view of water tank configuration Sicily/North Africa.

mules.jpg – 3 mules in a CG-4A; Burma.

VooDooTowLineTX.jpg – placard of First Transatlantic Glider Flight, 1 July 1943.

a_xcg10_arena1.jpg – Photo 124887 XCG-10 S/N 261099 Front view, St. Louis Arena.

a_xcg10_arena2TXT.jpg – Photo 124894 XCG-10 S/N261099 Top view St. Louis Arena.

xcg10_Draw.jpg – three view sketches of Laister-Kauffman 30 place model XCG-10.

XCG-10_Specs.jpg – AAF Glider Model XCG-10 performance specifications.

Tow_Release_Draw.jpg – Assembly blueprint /R.W. Huzzard.

ABW___AF.jpg – Photo 184741 XCG-10A S/N 261100 side view.

ABWA__JL.jpg – XCG-10A S/N 261099 rear quarter view New Orleans.

ABWB__HS.jpg – XCG-10A jeeps unloading; XCG-10A crew.

af10adat.jpg – Aberdeen performance graph CG-10A and other cargo gliders, 17 November 1944.

XCG-10A_DRAW.jpg – 3 view sketches of Laister-Kauffman XCG-10A, R.W. Huzzard.

XCG-10A_Spec1.jpg – AAF Glider Model XCG-10A specifications, 20 December 1943.

XCG-10A_Spec2.jpg – AAF Glider Model XCG-10A specifications, continued.

XCG-10A_Spec3.jpg – WF-5-16-44-2750-(45) XCG-10A description.

a_ycg10aTXT.jpg – YCG-10A S/N 45-44451 rear quarter view, circa 1946.

Mod80_L.jpg – AN 09-1-14 Three-Quarter cutaway of Pick-Up Unit, left half.

Mod80_R.jpg – AN 09-1-14 Three-Quarter cutaway of Pick-Up Unit, right half.

Winch_160_B17F_Draw.jpg – 2 view sketch Model 160 unit in B-17F /R.W. Huzzard.

Winch_160_Specs.jpg – Glider Pick-Up Unit Model 160X specifications, 21 December 1943.

DVD

Lee Jett piloting B-17, CG-4A/CG-13A/CG-10A snatch pickups Greenville SC, 3:31 min, Spring 1946.

LEE JETT COLLECTION

Electronic Images

Photo01.tif to Photo11.tif – CG-13 snatch pickup sequence in Troy OH.

Photo12.tif – CG-4A ambulance interior.

Photo13.tif and Photo14.tif – XCG-10 landing and takeoff.

Photo15.tif – CG-13 snatch.

Photo16.tif and Photo17.tif – Sequence of B23 snatching CG-4A.

Photo19.tif to Photo27.tif – CG-4A muddy field snatch recovery sequence.

Photo28.tif and Photo31.tif – Tug snatch equipments and crew.

Photo32.tif to Photo39.tif – Double snatch sequence.

Photo40.tif and Photo41.tif – C-47 snatch CG-4A *Open House Patterson Field*, 13 August 1944.

Photo42.tif and Photo43 – *"Airborne Attack" War Bond Demonstration*, C-46 towing CG-10; both have paratroopers to deploy.

LEON B. SPENCER COLLECTION

Printed Extracts

Pilot Training Manual for the CG-4A; AAF Manual No. 50, 17 March 1945.

Pilot Training Manual for the Skytrain C-47; AAF Manual No. 51-129, 2 September 1945.

G. ROBERT VEAZEY COLLECTION

Electronic Files

Air Pickup.pdf, *Air Pick-Up Handbook*, 1947.

Model 80H Spec Sheet.pdf

STARS.pdf, *Surface To Air Recovery System STARS,* Friends Journal, Summer 1996.

Printed Articles

K-13 Remembered, Friends Journal, Vol. 16, No. 4, Winter 1993.

Pickup!! #1 Friends Journal, Vol. 16, No. 2, Summer 1993.

Pickup!! #2 *Action in WWII,* Friends Journal, Vol. 17, No. 1, Spring 1994.

Pickup!! #3 *Mid-Air Recovery,* Friends Journal, Vol. 17, No. 4, Winter 1994.

Pickup!! #4 *Aerospace Recovery ...,* Friends Journal, Vol. 18, No. 3, Fall 1995.

Pickup! Air-To-Air Man Recovery, Friends Journal, Vol. 23, No. 1, Spring 2000.

A Walk Through the Mud, Friends Journal, Vol. 26, No. 2, Summer 2003.

My One-Day Trip to Hawaii, Friends Journal, Vol. 27, No. 4, Winter 2004.

GEORGE I. THEIS COLLECTION

Electronic Images

powgp.jpg – Prisoners Of War by his CG-4A.

wesel1.bmp – *Prior To Takeoff from France, En Route to LZ "N" 24 Mar 1945*.

wesel2.bmp – LZ "N" 24 - 25 Mar 1945.

wesel3.bmp – *Evacuation Route 24 - 25 Mar 1945*.

wesel4.bmp – My Glider in *LZ "N" 24 Mar 1945.*

HAGLEY MUSEUM AND LIBRARY

Hagley Museum and Library
P.O. Box 3630
Wilmington, DE 19807-0630
POC: Marjorie McNinch mmcninch@hagley.org (302) 658-2400

Photocopy

List of Research and Development Navy projects with All American Airways since 1944.

Engineering logbook, Glen K. Mead, 28 June 1944 – 2 February 1945.

All American Engineering Company Records 1937-1975, online research catalog at http://www.hagley.org/.

AIR FORCE HISTORICAL RESEARCH AGENCY, MAXWELL AIR FORCE BASE, ALABAMA

HQ AFHRA/RSA
600 Chennault Circle
Maxwell AFB, AL 36112-6424
POC: Archie.DiFante@maxwell.af.mil DSN 493-2447

Listing of multiple abstracts available from 16-mm film.

Photocopy

Flight test report TSFTE-1976, Performance of CG-46A-CG-10A towplane glider combination, 29 April 46. IRIS Number: 1040842.

Flight test report TSFTE-2051, Longitudinal stability of YCG-10A glider, 5 February 1947. IRIS Number: 1040852.

Aircraft assignment cards CG-10A 42-61099, 42-61100, 45-44450, 45-44451, 45-44452.

AAF Form No.14 Report of Major Accident, C-47A, 9 December 1948.

AAF Form No.14 Report of Major Accident, SB-17G, 13 December 1948.

Greenland Icecap Rescue files: miscellaneous message traffic.

Gliders get new pickup system by Sidney Peterson, Plane Facts, Volume 2, No. 5; May 1944. IRIS Number: 144882.

NATIONAL MUSEUM OF THE UNITED STATES AIR FORCE, WRIGHT-PATTERSON AFB, OH

National Museum of the United States Air Force Research Division/MUA
1100 Spaatz St.
Wright-Patterson AFB, OH 45433-7102
POC: Brett.Stolle@wpafb.af.mil

Electronic Images

CG-4A (01).jpg – muddy plain with snatched CG-4A.

CG-4A (02).jpg – runway snatch upon contact.

CG-10A (03).jpg – Jeep inside CG-10.

Photocopy Articles

CG-10A, Soaring, January-February 1945.

Rise and Fall of the Cargo Glider, Air Classics, October 1991.

Photocopy Excerpts

Pilot Training Manual for the CG-4A by Headquarters AAF.

Glider Tactics and Technique, Commanding General, Army Air Force, 24 January 1944.

Development of Gliders in the Army Air Force, 22 May 1945.

Pilot's Handbook for Army Model CG-10A Glider, 24 September 1945.

Erection and Maintenance Instructions for Model YCG-10A Gliders, 19 April 1946.

SILENT WINGS MUSEUM, LUBBOCK TX

Silent Wings Museum
6202 N I-27
Lubbock, TX 79403-9710
POC: (806) 775-2047
http://www.silentwingsmuseum.com/Visit%20The%20Museum/Research/articles.htm

Electronic Files

26th Mobile Reclamation and Repair Squadron.pdf, by Leon B. Spencer

The Wizards of Crookham Common.pdf, by Leon B. Spencer

Glider Tow Ropes WWII Military Gliders.pdf, by Leon B. Spencer

WWII USAAF Glider Aerial Retrieval System.pdf, by Leon B. Spencer

Cargo Glider Deceleration Parachute.pdf, *Development and Use...* by Leon B. Spencer

*The XCG17_Glider*s_Pics.doc.pdf, by Charles L. Day

VOO DOO was my name.doc.pdf, by Charles L. Day

Lowden Scans.pdf - *A Brief History of the Combat Glider in WWII* by John L. Lowden

DVD

AAF Training Film TF-1-3399 *Glider Snatch Pickup by C-47's*, 17:00 min TR-35.

SMITHSONIAN INSTITUTION NATIONAL AIR AND SPACE MUSEUM

Smithsonian Institution National Air and Space Museum
Archives Division
PO Box 37012
Room 3100, MRC 322
Washington, DC 20013-7012
POC: Joe Pruden, nasmrefdesk@si.edu
(202) 633-2320

Photocopy

Pilot's Manual for Army Model CG-10A Glider AN 09-15AB-1, 24 September 1945, CM-0017787.

Army Air Forces Technical Report: Test of 1/32-Scale Wind Tunnel Model of Laister-Kauffman XCG-10 Glider (Five-Foot Wind Tunnel Test No. 366) 28 May 1943. AL-113170-01.

Operational and Tactical Suitability of the XCG-10A Glider, 12 February 1945. AL-113172-1.

Laister-Kaufman LK-10 Glider pickup by Aircraft; Stinson SR-10. Image 89-153.

Stinson XC-81 (installed glider pickup unit). Image 7A40388.

Waco CG-4A Hadrian; Douglas C-47 Skytrain (glider pickup) Floyd Bennett Field, NY, circa 1946. Image 00128839.

Waco CG-4A Hadrian; Douglas B-23 Dragon, aerial pickup, Image USAF-A7644AC.

Laister-Kaufmann CG-10A General Assembly [Three View] *"Too Large to Duplicate, 42x56"* DD-0008465.

Microfilm

YCG-10A blueprints, *Inactive Contractor's Engineering Drawings & Data (Laister-Kauffmann)*, EDM 462c rolls A-F (2,935 frames), December 1954. DM-0000465.

www.ingramcontent.com/pod-product-compliance
Lightning Source LLC
Chambersburg PA
CBHW080826180526
45168CB00006B/2587